Design and Analysis of Experiments by Douglas Montgomery:
A Supplement for Using JMP®

Heath Rushing, Andrew Karl, and James Wisnowski

support.sas.com/bookstore

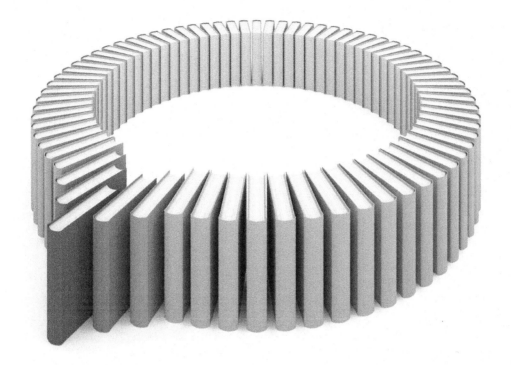

Gain Greater Insight into Your JMP® Software with SAS Books.

Discover all that you need on your journey to knowledge and empowerment.

 support.sas.com/bookstore
for additional books and resources.

THE POWER TO KNOW®

Contents

About This Book

Purpose

This is a supplement to the college textbook, *Design and Analysis of Experiments* by Douglas Montgomery using JMP software Version 10. Being that it is a supplement to an existing course textbook, this book is purely example-driven. This supplement demonstrates all examples from the course text, showing the reader how to complete the examples using JMP software Version 10. The following JMP platforms are used to produce examples: Custom Design, Screening Design, Response Surface Design, Mixture Design, Distribution, Fit Y by X, Matched Pairs, Fit Model, and Profiler.

Is This Book for You?

Although the majority of readers of this book are in academia (graduate or undergraduate programs), we anticipate that the other major audience includes scientists and engineers from the following industries: semiconductor, pharma/biopharma/medical device, chemical processing, manufacturing, and consumer goods. Level: Intermediate.

Prerequisites

At a minimum a reader of this book should have at least one introductory level course on statistics. However, this book is also appropriate for graduate level (Master's) students who have had advanced courses in statistics and even college professors (since it is a JMP software supplement).

Software Used to Develop This Book's Content

This book was developed using JMP Software Version 10.

JMP Data Tables and JMP Scripts

You can access the JMP data tables and JMP scripts for this book by linking to its author pages at http://support.sas.com/publishing/authors. Select the name of the author, then look for the cover image of this book and select "Example Code and Data" to display the JMP data tables and JMP scripts included in this book.

Output and Graphics

The output and graphics in this book were developed using JMP Software Version 10.

Author Pages for This Book

You can link to the author pages for this book at http://support.sas.com/publishing/authors. There you can find example code and data, free chapters, read the latest reviews, get updates, and more.

For an alphabetical listing of all the books for which example code and data is available, see http://support.sas.com/bookcode. Select a title to display the book's example code.

Additional Resources

SAS offers you a rich variety of resources to help build your SAS skills and explore and apply the full power of SAS software. Whether you are in a professional or academic setting, we have learning products that can help you maximize your investment in SAS.

Bookstore	http://support.sas.com/bookstore/
Training	http://support.sas.com/training/
Certification	http://support.sas.com/certify/
SAS Global Academic Program	http://support.sas.com/learn/ap/
SAS OnDemand	http://support.sas.com/learn/ondemand/
Knowledge Base	http://support.sas.com/resources/
Support	http://support.sas.com/techsup/
Training and Bookstore	http://support.sas.com/learn/
Community	http://support.sas.com/community/

Keep in Touch

We look forward to hearing from you. We invite questions, comments, and concerns. If you want to contact us about a specific book, please include the book title in your correspondence.

To Contact the Authors through SAS Press

By e-mail: saspress@sas.com

Via the Web: http://support.sas.com/author_feedback

SAS Books

For a complete list of books available through SAS visit http://support.sas.com/bookstore.

Phone: 1-800-727-3228

Fax: 1-919-677-8166

E-mail: sasbook@sas.com

SAS Book Report

Receive up-to-date information about all new SAS publications via e-mail by subscribing to the SAS Book Report monthly eNewsletter. Visit http://support.sas.com/sbr.

About These Authors

 Heath Rushing, Principal Consultant and co-founder of Adsurgo, LLC, an analytics consulting company that specializes in commercial and government training. Heath is a former professor from the Air Force Academy. He holds an M.S. degree in Operations Research from the Air Force Institute of Technology and has used JMP since 2001. After teaching at the Academy, Heath was a quality engineer and Six Sigma Black Belt in both biopharmaceutical manufacturing and Research and Development, where he used JMP to design and deliver training material in Six Sigma, Statistical Process Control (SPC), Design of Experiments (DOE), and Measurement Systems Analysis (MSA). In addition, Heath has been a symposium speaker at both national and international pharma and medical device conferences. Heath is an American Society of Quality (ASQ) Certified Quality Engineer and teaches JMP courses regularly, including a course he recently developed on Quality by Design (QbD) using JMP.

 Andrew T. Karl is a senior management consultant for Adsurgo, LLC, developing and teaching courses on a variety of statistical topics for the U.S. Department of Defense, Fortune 500 corporations, and international clients. He received his B.A. in mathematics from the University of Notre Dame and his Ph.D. in statistics from Arizona State University. Dr. Karl's research interests focus on computation and applications of non-nested linear and nonlinear mixed models, including value-added problems. Additionally, he frequently works with problems from data mining, text mining, and experimental design.

 Jim Wisnowski is co-founder and principal at Adsurgo, LLC, an analytics consulting company that specializes in commercial and government training. He has a Ph.D. in Industrial Engineering from Arizona State University and has published numerous journal articles and textbook chapters. He was an award-winning professor and statistics chair while at the United States Air Force Academy. Jim retired from the Air Force, where he held various analytical and leadership positions throughout the Department of Defense in training, test and evaluation, human resources, logistics, systems engineering, and acquisition.

Learn more about these authors by visiting their author pages where you can download free chapters, read the latest reviews, get updates, and more:

- http://support.sas.com/rushing
- http://support.sas.com/wisnowski
- http://support.sas.com/karl

Acknowledgments

All three authors have been students of Dr. Douglas Montgomery; two literally, one figuratively. We could not have even considered writing this book without the experiences of having studied under Dr. Montgomery, and both taking and teaching courses using multiple versions of his text. It is generally regarded as the most useful text for design of experiments; we understand why.

We would like to acknowledge the contributions of those who helped us write this supplement. Thanks to the several technical reviewers at SAS who took the time to carefully review the draft: Mark Bailey, Mia Stephens, Paul Marovich, and Di Michelson. Also, thanks to Dr. Jianbiao John Pan from California Polytechnic State University San Luis Obispo who provided comments and suggestions for the supplement. We would be remiss if we did not acknowledge the contributions of the JMP software development team, specifically Bradley Jones and Chris Gotwalt, who continually improve the software and listen to customer input.

Thank you to Shelley Sessoms, the SAS Press Acquisitions Editor, who provided us with this opportunity. Thanks to the SAS Publications Editing and Production staff for making corrections and improvements to the book: Kathy Underwood, Candy Farrell, Thais Fox, Robert Harris, Stacy Suggs, and Denise Jones. Finally, a special thanks to SAS Press Developmental Editor John West and SAS Publications Marketing Specialist Cindy Puryear for managing this project and keeping us on track.

Introduction

The analysis of a complex process requires the identification of target quality attributes that characterize the output of the process and of factors that may be related to those attributes. Once a list of potential factors is identified from subject-matter expertise, the strengths of the associations between those factors and the target attributes need to be quantified. A naïve, one-factor-at-a-time analysis would require many more trials than necessary. Additionally, it would not yield information about whether the relationship between a factor and the target depends on the values of other factors (commonly referred to as interaction effects between factors). As demonstrated in Douglas Montgomery's *Design and Analysis of Experiments* textbook, principles of statistical theory, linear algebra, and analysis guide the development of efficient experimental designs for factor settings. Once a subset of important factors has been isolated, subsequent experimentation can determine the settings of those factors that will optimize the target quality attributes. Fortunately, modern software has taken advantage of the advanced theory. This software now facilitates the development of good design and makes solid analysis more accessible to those with a minimal statistical background.

Designing experiments with specialized design of experiments (DOE) software is more efficient, complete, insightful, and less error-prone than producing the same design by hand with tables. In addition, it provides the ability to generate algorithmic designs (according to one of several possible optimality criteria) that are frequently required to accommodate constraints commonly encountered in practice. Once an experiment has been designed and executed, the analysis of the results should respect the assumptions made during the design process. For example, split-plot experiments with hard-to-change factors should be analyzed as such; the constraints of a mixture design must be incorporated; non-normal responses should either be transformed or modeled with a generalized linear model; correlation between repeated observations on an experimental unit may be modeled with random effects; non-constant variance in the response variable across the design factors may be modeled, etc. Software for analyzing designed experiments should provide all of these capabilities in an accessible interface.

JMP offers an outstanding software solution for both designing and analyzing experiments. In terms of design, all of the classic designs that are presented in the textbook may be created in JMP. Optimal designs are available from the JMP Custom Design platform. These designs are extremely useful for cases where a constrained design space or a restriction on the number of experimental runs eliminates classical designs from consideration. Multiple designs may be created and compared with methods described in the textbook, including the Fraction of Design Space plot. Once a design is chosen, JMP will randomize the run order and produce a data table, which the researcher may use to store results. Metadata for the experimental factors and response variables is attached to the data table, which simplifies the analysis of these results.

The impressive graphical analysis functionality of JMP accelerates the discovery process particularly well with the dynamic and interactive profilers and plots. If labels for plotted points overlap, can by clicking and dragging the labels. Selecting points in a plot produced from a table selects the appropriate rows in the table and highlights the points corresponding to those rows in all other graphs produced from the table. Plots can be shifted and rescaled by clicking and dragging the axes. In many other software packages, these changes are either unavailable or require regenerating the graphical output.

An additional benefit of JMP is the ease with which it permits users to manipulate data tables. Data table operations such as sub-setting, joining, and concatenating are available via intuitive graphical interfaces. The relatively short learning curve for data table manipulation enables new users to prepare their data without remembering an extensive syntax. Although no command-line knowledge is necessary, the underlying JMP scripting language (JSL) scripts for data manipulation (and any other JMP procedure) may be saved and edited to repeat the analysis in the future or to combine with other scripts to automate a process.

This supplement to *Design and Analysis of Experiments* follows the chapter topics of the textbook and provides complete instructions and useful screenshots to use JMP to solve every example problem. As might be expected, there are often multiple ways to perform the same operation within JMP. In many of these cases, the different possibilities are illustrated across different examples involving the relevant operation. Some theoretical results are discussed in this supplement, but the emphasis is on the practical application of the methods developed in the textbook. The JMP DOE functionality detailed here represents a fraction of the software's features for not only DOE, but also for most other areas of applied statistics. The platforms for reliability and survival, quality and process control, time series, multivariate methods, and nonlinear analysis procedures are beyond the scope of this supplement.

Simple Comparative Experiments

The problem of testing the effect of a single experimental factor with only two levels provides a useful introduction to the statistical techniques that will later be generalized for the analysis of more complex experimental designs. In this chapter, we develop techniques that will allow us to determine the level of statistical significance associated with the difference in the mean responses of two treatment levels. Rather than only considering the difference between the mean responses across the treatments, we also consider the variation in the responses and the number of runs performed in the experiment. Using a t-test, we are able to quantify the likelihood (expressed as a p-value) that the observed treatment effect is merely noise. A "small" p-value (typically taken to be one smaller than $\alpha = 0.05$) suggests that the observed data are not likely to have occurred if the null hypothesis (of no treatment effect) were true.

A related question involves the likelihood that the null hypothesis is rejected given that it is false (the power of the test). Given a fixed significance level, α (our definition of what constitutes a "small" p-value), theorized values for the pooled standard deviation, and a minimum threshold difference in treatment means, it is possible to solve for the minimum sample size that is necessary to achieve a desired power. This procedure is useful for determining the number of runs that must be included in a designed experiment.

In the first example presented in this chapter, a scientist has developed a modified cement mortar formulation that has a shorter cure time than the unmodified formulation. The scientist would like to test if the modification has affected the bond strength of the mortar. To study whether the two formulations, on average, produce bonds of different strengths, a two-sided t-test is used to analyze the observations from a randomized experiment with 10 measurements from each formulation. The null hypothesis of this test is that the mean bond strengths produced by the two formulations are equal; the alternative hypothesis is that mean bond strengths are not equal.

We also consider the advantages of a paired t-test, which provides an introduction to the notion of blocking. This test is demonstrated using data from an experiment to test for similar performance of two different tips that are placed on a rod in a machine and pressed into metal test coupons. A fixed pressure is applied to the tip, and the depth of the resulting depression is measured. A completely randomized design would apply the tips in a random order to the test coupons (making only one measurement on each coupon). While this design would produce valid results, the power of the test could be increased by removing noise from the coupon-to-coupon variation. This may be achieved by applying both tips to each coupon (in a random order) and measuring the difference in the depth of the depressions. A one-sample t-test is then used for the null hypothesis that the mean difference across the coupons is equal to 0. This procedure reduces experimental error by eliminating a noise factor.

This chapter also includes an example of procedures for testing the equality of treatment variances, and a demonstration of the t-test in the presence of potentially unequal group variances. This final test is still valid when the group variances are equal, but it is not as powerful as the pooled t-test in such situations.

Section 2.2 Basic Statistical Concepts

1. Open Tension-Bond.jmp.

2. Select **Analyze > Distribution**.

3. Select *Strength* for **Y, Columns.**

4. Select *Mortar* for **By**. As we will see in later chapters, these fields will be automatically populated for data tables that were created in JMP.

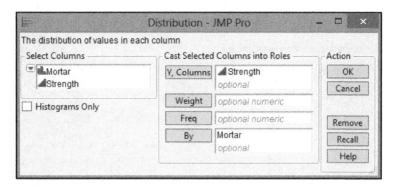

5. Click **OK**.

6. Click the red triangle next to Distributions Mortar=Modified and select **Uniform Scaling**.

7. Repeat step 6 for Distributions Mortar=Unmodified.

8. Click the red triangle next to Distributions Mortar=Modified and select **Stack**.

9. Repeat step 8 for Distributions Mortar=Unmodified.

10. Hold down the *Ctrl* key and click the red triangle next to Strength. Select **Histogram Options > Show Counts**. Holding down *Ctrl* applies the command to all of the histograms created by the Distribution platform; it essentially "broadcasts" the command.

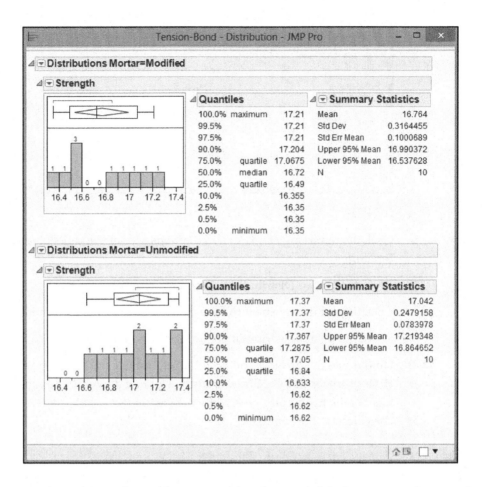

It appears from the overlapped histograms that the unmodified mortar tends to produce stronger bonds than the modified mortar. The unmodified mortar has a mean strength of 17.04 kgf/cm² with a standard deviation of 0.25 kgf/cm². The modified mortar has a mean strength of 16.76 kgf/cm² with a standard deviation of 0.32 kgf/cm². A naïve comparison of mean strength indicates that the unmodified mortar outperforms the modified mortar. However, the difference in means could simply be a result of sampling fluctuation. Using statistical theory, our goal is to incorporate the sample standard deviations (and sample sizes) to quantify how likely it is that the difference in mean strengths is due only to sampling error. If it turns out to be unlikely, we will conclude that a true difference exists between the mortar strengths.

11. Select **Analyze > Fit Y by X**.

12. Select *Strength* for **Y, Response** and *Mortar* for **X, Grouping**.

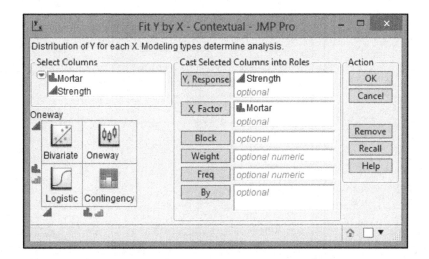

The Fit Y by X platform recognizes this as a one-way ANOVA since the response, *Strength*, is a continuous factor, and the factor *Mortar* is a nominal factor. When JMP is used to create experimental designs, it assigns the appropriate variable type to each column. For imported data, JMP assigns a modeling type—continuous ▰, ordinal ▰, or nominal ▰—to each variable based on attributes of that variable. A different modeling type may be specified by right-clicking the modeling type icon next to a column name and selecting the new type.

13. Click **OK**.

14. To create box plots, click the red triangle next to One-way Analysis of Strength by Mortar and select **Quantiles**.

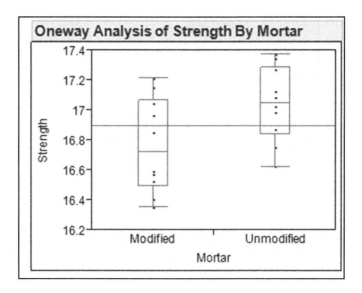

The median modified mortar strength (represented by the line in the middle of the box) is lower than the median unmodified mortar strength. The similar length of the two boxes (representing the interquartile ranges) indicates that the two mortar formulations result in approximately the same variability in strength.

15. Keep the Fit Y by X platform open for the next exercise.

Section 2.4.1 Hypothesis Testing

1. Return to the Fit Y by X platform from the previous exercise.

2. Click the red triangle next to One-way Analysis of Strength by Mortar and select **Means/Anova/Pooled t**.

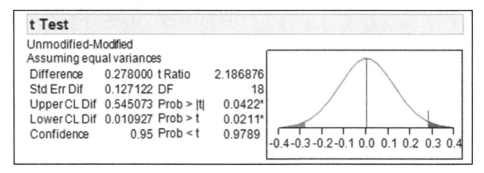

The **t-test** report shows the two-sample t-test assuming equal variances. Since we have a two-sided alternative hypothesis, we are concerned with the p-value labeled Prob > |t| = 0.0422. Since we have set $\alpha=0.05$, we reject the null hypothesis that the mean strengths produced by the two formulations of mortar are equal and conclude that the mean strength of the modified mortar and the mean strength of the unmodified mortar are (statistically) significantly different. In practice, our next step would be to decide from a subject-matter perspective if the difference is practically significant.

Before accepting the conclusion of the t test, we should use diagnostics to check the validity of assumptions made by the model. Although this step is not shown for every example in the text, it is an essential part of every analysis. For example, a quantile plot may be used to check the assumptions of normality and identical population variances. Though not shown here, a plot of the residuals against run order could help identify potential violations of the assumed independence across runs (the most important of the three assumptions).

3. Click the red triangle next to One-way Analysis of Strength by Mortar and select **Normal Quantile Plot > Plot Quantile by Actual**.

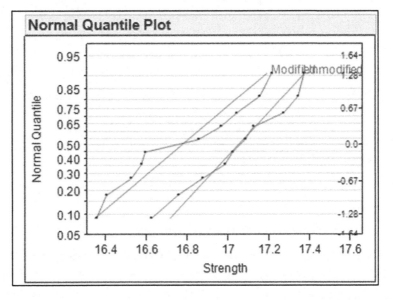

The points fall reasonably close to straight lines in the plot, suggesting that the assumption of normality is reasonable. The slopes of the lines are proportional to the standard deviations in each comparison group. These slopes appear to be similar, supporting the decision to assume equal population variances.

4. Select **Window > Close All**.

Section 2.4.3 Choice of Sample Size

1. To determine the necessary sample size for a proposed experiment, select **DOE > Sample Size and Power**.

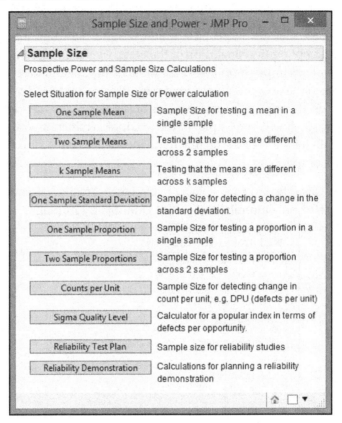

2. Click **Two Sample Means**.

3. Enter 0.25 for **Std Dev**, 0.5 for **Difference to detect**, and 0.95 in **Power**. Notice that the **Difference to detect** requested here is the actual difference between group means, not the scaled difference, δ, described in the textbook.

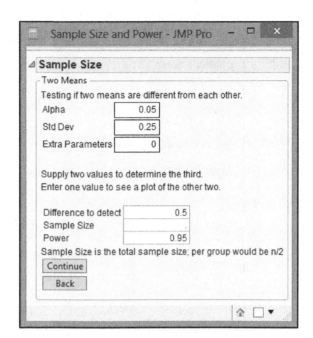

4. Click **Continue**. A value of 16 then appears in **Sample Size**. Thus, we should allocate 8 observations to each treatment (n1 = n2 = 8).

5. Suppose we use a sample size of n1 = n2 = 10. What is the power for detecting difference of 0.25 kgf/cm²? Delete the value 0.95 from the **Power** field, change **Difference to detect** to 0.25, and set **Sample Size** to 20.

6. Click **Continue**.

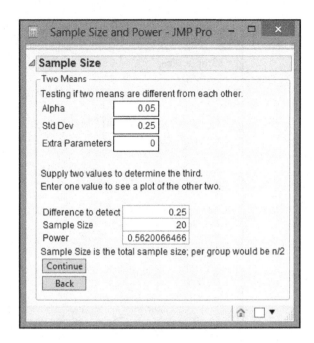

The power has dropped to 0.56. That is, if the model assumptions hold and the true pooled standard deviation is 0.25, only 56% of the experiments (assuming that we repeat this experiment several times) with 10 measurements from each group would successfully detect the difference of 0.25 kgf/cm². What sample size would be necessary to achieve a power of 0.9 for this specific difference to detect?

7. Clear the **Sample Size** field and enter 0.9 for **Power**.

8. Click **Continue**.

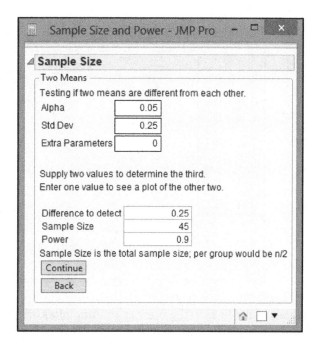

The required total sample size is 45. This means that we need at least 22.5 observations per group. Rounding up, we see that we need at least 23 observations from each group to achieve a power of at least 0.9. We could have left the **Power** field blank, specifying only that the **Difference to detect** is 0.25. The Sample Size and Power platform would then have produced a power curve, displaying **Power** as a function of **Sample Size**.

9. Select **Window > Close All**.

Example 2.1 Hypothesis Testing

1. Open Fluorescence.jmp.

2. Click **Analyze > Fit Y by X**.

3. Select *Fluorescence* for **Y, Response** and *Tissue* for **X, Factor**.

4. Click **OK**.

5. Click the red triangle next to One-way Analysis of Fluorescence by Tissue and select **Normal Quantile Plot > Plot Quantile by Actual**.

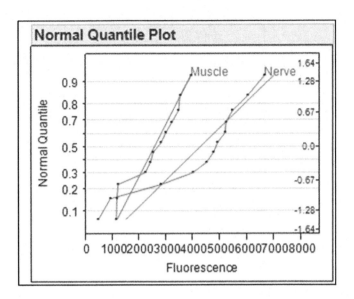

Since the slopes of the lines in the normal quantile plots are proportional to the sample standard deviations of the treatments, the difference between the slopes of the lines for *Muscle* and *Nerve* indicates that the variances may be different between the groups. As a result, we will use a form of the t-test that does not assume that the population variances are equal. Formal testing for the equality of the treatment variances is illustrated in Example 2.3 at the end of this chapter.

6. Click the red triangle next to One-way Analysis of Fluorescence by Tissue and select **t-Test.**

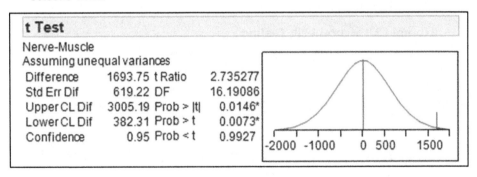

The p-value for the one-sided hypothesis test is 0.0073, which is less than the set α of 0.05. We therefore reject the null hypothesis and conclude that the mean normalized fluorescence for nerve tissue is greater than the mean normalized fluorescence for muscle tissue. Subject matter knowledge would need to determine if there is a practical difference; confidence intervals for the differences (reported in JMP) can be beneficial for this assessment.

7. Select **Window > Close All**.

Section 2.5.1 The Paired Comparison Problem

1. Open Hardness-Testing.jmp

2. Select **Analyze > Matched Pairs**.

3. Select *Tip 1* and *Tip 2* for **Y, Paired Response**.

4. Click **OK**.

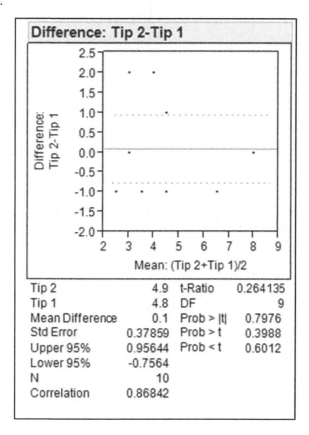

Tip 2	4.9	t-Ratio	0.264135
Tip 1	4.8	DF	9
Mean Difference	0.1	Prob > |t|	0.7976
Std Error	0.37859	Prob > t	0.3988
Upper 95%	0.95644	Prob < t	0.6012
Lower 95%	-0.7564		
N	10		
Correlation	0.86842		

The p-value, Prob > |t| = 0.7976, indicates that there is no evidence of a difference in the performance of the two tips. This p-value is larger than the standard significance level of $\alpha = 0.05$.

5. Leave Hardness-Testing.jmp open for the next exercise.

Section 2.5.2 Advantages of the Paired Comparison Design

1. Return to the Hardness-Testing table opened in the previous example.

2. Select **Tables > Stack**. This will create a file in long format with one observation per row. Most JMP platforms expect data to appear in long format.

3. Select *Tip 1* and *Tip 2* for **Stack Columns**.

4. Type "Depth" in the **Stacked Data Column** field.

5. Type "Tip" in the **Source Label Column** field.

6. Type "Hardness-Stacked" in the **Output table name** field.

7. Click **OK**.

8. Hardness-Stacked is now the current data table. Select **Analyze > Fit Y by X**.

9. Select *Depth* for **Y, Response** and *Tip* for **X, Grouping**.

10. Click **OK**.

11. Click the red triangle next to One-way Analysis of Depth by Tip and select **Means/Anova/Pooled t**.

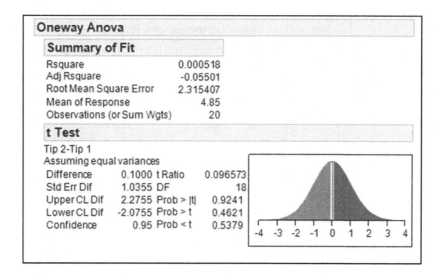

The root mean square error of 2.315407 is the pooled standard deviation estimate from the t-test. Compared to the standard deviation estimate of 1.20 from the paired difference test, we see that blocking has reduced the estimate of variability considerably. Though we do not work through the details here, it would be possible to perform this same comparison for the Fluorescence data from Example 2.1.

12. Leave Hardness-Stacked.jmp and the Fit Y by X output window open for the next exercise.

Example 2.3 Testing for the Equality of Variances

This example demonstrates how to test for the equality of two population variances. Section 2.6 of the textbook also discusses hypothesis testing for whether the variance of a single population is equal to a given constant. Though not shown here, the testing for a single variance may be performed in the Distribution platform.

1. Return to the Fit Y by X platform from the previous example.

2. Click the red triangle next to One-way Analysis of Depth by Tip and select **Unequal Variances**.

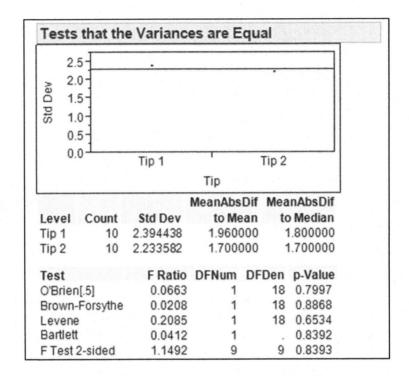

Tests that the Variances are Equal

Level	Count	Std Dev	MeanAbsDif to Mean	MeanAbsDif to Median
Tip 1	10	2.394438	1.960000	1.800000
Tip 2	10	2.233582	1.700000	1.700000

Test	F Ratio	DFNum	DFDen	p-Value
O'Brien[.5]	0.0663	1	18	0.7997
Brown-Forsythe	0.0208	1	18	0.8868
Levene	0.2085	1	18	0.6534
Bartlett	0.0412	1	.	0.8392
F Test 2-sided	1.1492	9	9	0.8393

3. Save Hardness-Stacked.jmp.

The p-value for the F test (described in the textbook) for the null hypothesis of equal variances (with a two-sided alternative hypothesis) is 0.8393. The data do not indicate a difference with respect to the variances of depth produced from Tip 1 versus Tip 2. Due to the use of a slightly different data set, the F Ratio of 1.1492 reported here is different from the ratio of 1.34 that appears in the book. Furthermore, the textbook uses a one-sided test with an alternative hypothesis. That hypothesis is that the variance of the depth produced by Tip 1 is greater than that produced by Tip 2. Since the sample standard deviation from Tip 1 is greater than that from Tip 2, the F Ratios for the one- and two-sided tests are both equal to 1.1492, but the p-value for the one-sided test would be 0.4197.

It is important to remember that the F test is extremely sensitive to the assumption of normality. If the population has heavier tails than a normal distribution, this test will reject the null hypothesis (that the population variances are equal) more often than it should. By contrast, the Levene test is robust to departures from normality.

4. Select **Window > Close All**.

Experiments with a Single Factor: The Analysis of Variance

In this chapter, the t-test is generalized to accommodate factors with more than two levels. The method of analysis of variance (ANOVA) introduced here allows us to study the equality of the means of three or more factor levels. ANOVA partitions the total sample variance into two parts: the variance explained by the factor under study, and the remaining, unexplained variance.

The method makes several assumptions about the distribution of the random error term in the model. If the model structure represents the true structure of the process, the model residuals may be thought of as random numbers generated from the distribution of the random error term, which is typically assumed to be a normal distribution. Several diagnostics are available for the residuals. They may be plotted on a normal quantile plot to check the assumption of normality of the random error term. They may also be plotted against the predicted values: the residuals and predicted values ought to be independent, and no patterns should be present in the plot. ANOVA also assumes that the error terms are independent and identically distributed. This chapter considers two formal tests, Bartlett's and Levene's, for the homogeneity of residual variance across factor levels. If any of the residual diagnostics show abnormalities, a transformation of the response variable is often useful for improving the model fit.

When the ANOVA test rejects the null hypothesis that all treatment means are equal, it is often necessary to know which factor levels are significantly different from each other. Special techniques are necessary for multiple comparisons of different linear combinations of factor level means in order to control the so-called experimentwise error rate. Examples are presented for Tukey's HSD (honestly significant difference) test and the Fisher (Student's t) least significant difference method. If one of the factors represents a control group, Dunnett's test may be used to compare the control group with each of the other factor levels.

Other topics covered include power analysis for ANOVA to determine a required sample size, an introduction to the random effects models that are useful when the factor levels are only a sample of a larger population, and an example of a nonparametric method. The Kruskal-Wallis test relaxes the assumption that the response distribution is normal in each factor level, though it does require that the distributions across factor levels have the same shape.

The first example will illustrate how to build an ANOVA model from data imported into JMP. This entails specifying the response column, the factor column, and ensuring that the factor column is set to the nominal modeling type. Afterward, we will show how models may be designed in JMP, and how the appropriate modeling options are saved as scripts attached to the data table. For the remainder of the text, we will assume that the data tables are created in JMP.

Section 3.1 A One-way ANOVA Example

1. Open Etch-Rate-Import.jmp.

2. Click the blue icon (triangle) next to *Power* in the Columns panel and select **Nominal**.

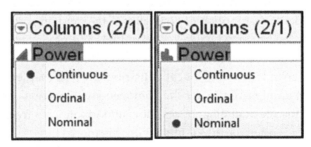

Even though the levels of *Power* are ordinal, we are not incorporating that information into the current analysis. This distinction is not critical since *Power* is only a factor, and not the response. Treating an ordinal factor as nominal yields the same model fit. For a response variable, a nomial modeling type prompts a multinomial logistic regression, while an ordinal modeling type prompts an ordered logistic regression.

3. Select **Analyze > Fit Y by X**.

4. Select *Etch Rate* and click **Y, Response**.

5. Select *Power* and click **X, Factor**.

6. Click **OK**.

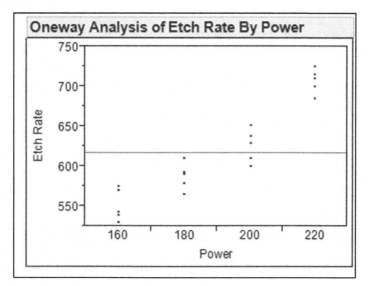

7. To produce a box plot, click the red triangle next to One-way Analysis of Etch Rate By Power and select **Quantiles**.

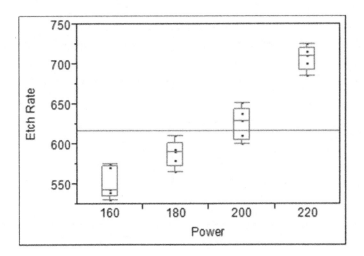

The box plots show that the etch rate increases as power increases, and that the variability of the etch rate is roughly the same for each power setting.

8. Select **Window > Close All**. We will now demonstrate how this model can be created in JMP.

9. Select **DOE > Full Factorial Design.**

10. Under **Response Name**, double click **Y** and change the response name to *Etch Rate.*

11. In the **Factors** section, select **Categorical > 4 Level.**

12. Double-click the name of the new factor, *X1*, and change it to *Power.*

13. Likewise, change the **Values** of the new factor from *L1, L2, L3,* and *L4* to *160, 180, 200,* and *220,* respectively.

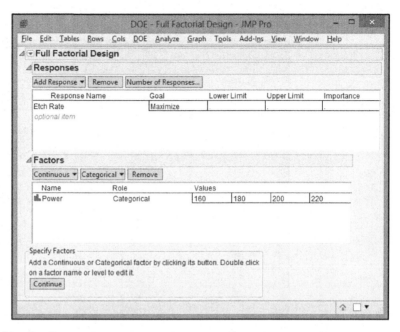

14. Click **Continue**.

15. Leave **Run Order** set to **Randomize**. Then, the experiment should be run in the order in which the rows appear in the resulting JMP table.

Number of Runs: 4 indicates that the design requires four different runs. Once **Make Table** has been clicked, **Number of Runs** will change to 20, reflecting the runs needed for the four replicates.

16. Enter *4* for **Number of Replicates**. This indicates the original 4 runs will be replicated 4 times for a total of 20 runs.

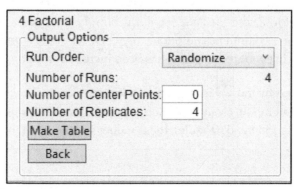

17. Click **Make Table**.

	Pattern	Power	Etch Rate
1	4	220	715
2	1	160	570
3	2	180	593
4	2	180	610
5	3	200	600
6	1	160	542
7	2	180	579
8	1	160	539
9	2	180	565
10	4	220	685
11	1	160	530
12	4	220	710
13	4	220	725
14	3	200	637
15	4	220	700
16	1	160	575
17	3	200	610
18	3	200	629
19	2	180	590
20	3	200	651

A new data table has been created with three columns. The *Pattern* column indicates which combination of factor levels are being used for the current row. Since there is only one factor, *Power*, the *Pattern* column simply indicates which level of *Power* is being run. The *Power* column has automatically been set to the **Nominal** modeling type. Additional metadata about the columns has been included from the Full Factorial platform, as indicated by the ✱ icons in the **Columns** section. The dots in the *Etch Rate* column represent missing values. As the experiments are conducted (in the randomized order presented in the data table), these values will be filled in by the engineer.

All of the JMP platforms demonstrated in this book are capable of fitting models in the presence of missing data. That is, if it is not possible to perform the 20th run, which is at the *Power* setting of 220, it would still be possible to analyze the first 19 runs. However, missing observations can affect the aliasing structure of a design, which will be discussed in later chapters. In addition, if the cause of the missing values is related to the response (missing not at random), then the resulting estimates could be biased.

18. Select **Window > Close All**.

19. Open Etch-Rate.jmp. This data table was created in JMP using the Full Factorial Design platform.

20. Select **Analyze > Fit Model**.

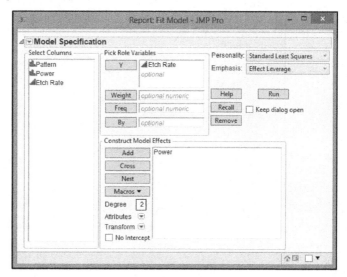

Etch Rate is automatically populated into the **Y** field, and the nominal *Power* factor is automatically added as a model effect. The corresponding fields in the Fit Y by X platform will not be automatically populated. However, the Fit Model platform is more general than the Fit Y by X platform and will be used much more frequently. Setting default column roles for the Fit Y by X and other platforms may be achieved via the **Cols > Preselect Role** menu. Please note that this can also be accomplished by selecting **Run Script** from the (red triangle associated with the) **Model** script contained in the Table Panel of the data table.

21. Select **Window > Close All**.

Example 3.1 The Plasma Etching Experiment

1. Open Etch-Rate.jmp.

2. Select **Analyze > Fit Y by X**.

3. Select *Etch Rate* and click **Y, Response**.

4. Select *Power* and click **X, Factor**.

5. Click **OK**.

6. Click the red triangle next to One-way Analysis of Etch Rate By Power and select **Means/Anova**.

Oneway Anova

Summary of Fit

Rsquare	0.92606
Adj Rsquare	0.912196
Root Mean Square Error	18.26746
Mean of Response	617.75
Observations (or Sum Wgts)	20

Analysis of Variance

Source	DF	Sum of Squares	Mean Square	F Ratio	Prob > F
Power	3	66870.550	22290.2	66.7971	<.0001*
Error	16	5339.200	333.7		
C. Total	19	72209.750			

Means for Oneway Anova

Level	Number	Mean	Std Error	Lower 95%	Upper 95%
160	5	551.200	8.1695	533.88	568.52
180	5	587.400	8.1695	570.08	604.72
200	5	625.400	8.1695	608.08	642.72
220	5	707.000	8.1695	689.68	724.32

Std Error uses a pooled estimate of error variance

The p-value for the F test of the null hypothesis of the equality of treatment means is <.0001. We conclude that the treatment means differ.

7. Leave the Etch-Rate data table open for the next exercise.

Example 3.3 Treatment Effects and Confidence Intervals.

1. Return to the Etch-Rate data table.

2. Click **Analyze > Fit Model**. As shown in the first example of this chapter, the modeling roles are pre-specified. Unless a different model needs to be fit than the one specified by the script attached to the data table, the screenshot of the Fit Model platform may be omitted.

3. Click **Run**. As noted previously, the Fit Model platform could also have been launched by clicking the red triangle next to the Model script in the Etch-Rate data table and clicking **Run Script**. This script was created by the Full Factorial Design platform and attached to the data table produced therein.

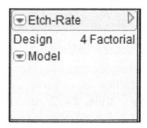

4. Scroll down to the Parameter Estimates report. It may be necessary to click the gray triangle next to the report title in order to expand the output window.

Parameter Estimates

Term	Estimate	Std Error	t Ratio	Prob>|t|
Intercept	617.75	4.084728	151.23	<.0001*
Power[160]	-66.55	7.074956	-9.41	<.0001*
Power[180]	-30.35	7.074956	-4.29	0.0006*
Power[200]	7.65	7.074956	1.08	0.2956

The coefficients for power settings of 160, 180, and 200 provided by JMP match those in the textbook. The intercept represents the grand mean of the observations.

5. To see the estimate for 220, click the red triangle next to Response Etch Rate and select **Estimates > Expanded Estimates**.

6. To display the confidence intervals for the parameter estimates, right-click inside the Expanded Estimates report, and select **Columns > Lower 95%** and **Columns > Upper 95%**.

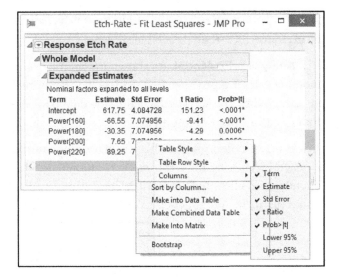

Alternatively, you can click the red triangle next to Response Etch Rate and select **Regression Reports > Show All Confidence Intervals.**

Expanded Estimates

Nominal factors expanded to all levels

Term	Estimate	Std Error	t Ratio	Prob>|t|	Lower 95%	Upper 95%
Intercept	617.75	4.084728	151.23	<.0001*	609.09076	626.40924
Power[160]	-66.55	7.074956	-9.41	<.0001*	-81.54824	-51.55176
Power[180]	-30.35	7.074956	-4.29	0.0006*	-45.34824	-15.35176
Power[200]	7.65	7.074956	1.08	0.2956	-7.348236	22.648236
Power[220]	89.25	7.074956	12.61	<.0001*	74.251764	104.24824

7. In addition to viewing the Expanded Estimates, you can also click the red triangle next to Response Etch Rate and select **Effect Screening > Scaled Estimates**.

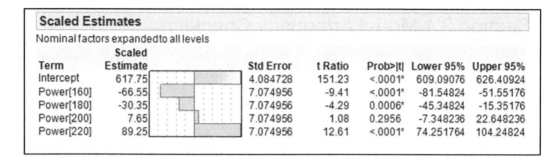

Scaled Estimates

Nominal factors expanded to all levels

Term	Scaled Estimate		Std Error	t Ratio	Prob>|t|	Lower 95%	Upper 95%
Intercept	617.75		4.084728	151.23	<.0001*	609.09076	626.40924
Power[160]	-66.55		7.074956	-9.41	<.0001*	-81.54824	-51.55176
Power[180]	-30.35		7.074956	-4.29	0.0006*	-45.34824	-15.35176
Power[200]	7.65		7.074956	1.08	0.2956	-7.348236	22.648236
Power[220]	89.25		7.074956	12.61	<.0001*	74.251764	104.24824

8. The Scaled Estimates report produces the same output as the Expanded Estimates report, in addition to a graphical representation of the magnitude of the treatment effects.

9. By clicking the red triangle next to Response Etch Rate and selecting **Factor Profiling > Profiler**, you obtain an interval plot of the mean responses and their confidence intervals.

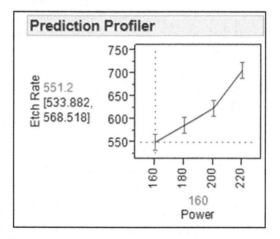

The Prediction Profiler provides the estimate mean response together with a confidence interval for each power setting. We will not explore the full functionality of the Prediction Profiler here, but it may be used for optimizing parameter settings to achieve a desired response.

10. Leave the Fit Model platform open for the next exercise.

Section 3.4 Model Adequacy Checking

1. In the Fit Model platform from the previous exercise, scroll down to the Residuals by Predicted plot. This plot is discussed in Section 3.4.3 of the textbook.

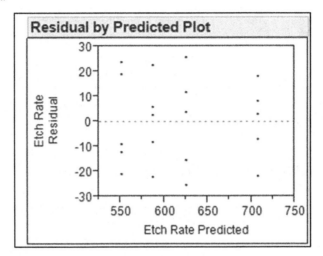

The variance appears to be constant across the range of predicted etch rates, and no patterns emerge from the plot. Because only a single categorical factor, *Power*, is included in the model, the validity of the assumption of constant residual variance may also be checked with formal tests, as described in Example 3.4.

2. To check the normality assumption, a quantile plot is commonly used. In JMP, the first step is to generate residuals for each observation. Click the red triangle next to Response Etch Rate and select **Save Columns > Residuals**.

3. Return to the Etch-Rate data table (you can use a shortcut by clicking the table icon ⊞ at the bottom of the report window) and notice the new column *Residual Etch Rate*.

	Power	Etch Rate	Residual Etch Rate
1	200	600	-25.4
2	220	725	18
3	220	700	-7
4	160	575	23.8
5	160	542	-9.2
6	180	565	-22.4

4. Click **Analyze > Distribution**.

5. Select *Residual Etch Rate* for **Y, Columns**.

6. Click **OK.**

7. Click the red triangle next to Residual Etch Rate and select **Continuous Fit > Normal**.

8. Scroll down and click the red triangle next to Fitted Normal and select **Diagnostic Plot**.

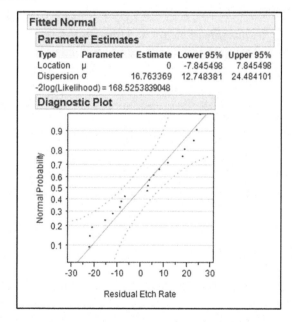

The error distribution appears to be approximately normal as the points fall relatively close to a straight line. We may also perform a Shapiro-Wilk test for the hypothesis that the residuals are from a normal distribution.

9. Click the red triangle next to Fitted Normal and select **Goodness of Fit**.

Goodness-of-Fit Test

Shapiro-Wilk W Test

W	Prob<W
0.937520	0.2152

Note: Ho = The data is from the Normal distribution. Small p-values reject Ho.

With a p-value of 0.2152, the residuals do not display a significant number of departures from normality.

10. Section 3.4.2 of the text discusses plotting the residuals in a time sequence to look for correlations between subsequent runs, which would represent a violation of the (important) independence assumption. To generate this plot, select **Analyze > Modeling > Time Series**.

11. Select *Residual Etch Rate* for **Y, Time Series**.

12. Click **OK**.

13. Click **OK** for the next dialog setting the number of autocorrelation lags to 19.

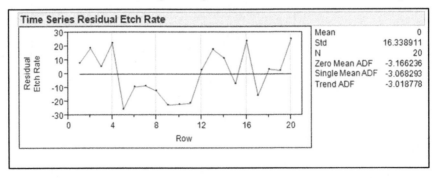

The first four residuals are all greater than zero while the next seven are all less than zero. There could be a systematic cause for this behavior, such as an omitted covariate (e.g. operator or ambient temperature). Though it is beyond the scope of our discussion, the **Time Series** platform may be used to detect correlations between subsequent runs. Furthermore, detecting patterns in a residual by time plot is analogous to detecting out-of-control conditions on a control chart (e.g. using the Western Electric Rules). If the residual by time plot signals as out of control according to these rules, it could indicate a shift in the behavior of the process during the course of the experiment.

14. Leave Etch-Rate open for the next exercise.

Example 3.4 Test for Equal Variances

1. Select **Analyze > Fit Y by X**.

2. Select *Etch Rate* and click **Y, Response**.

3. Select *Power* and click **X, Factor**.

4. Click **OK**.

5. Click the red arrow next to One-way Analysis of Etch Rate By Power and select **Unequal Variances**.

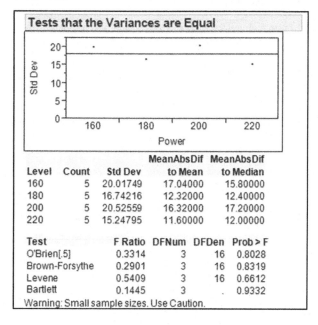

Level	Count	Std Dev	MeanAbsDif to Mean	MeanAbsDif to Median
160	5	20.01749	17.04000	15.80000
180	5	16.74216	12.32000	12.40000
200	5	20.52559	16.32000	17.20000
220	5	15.24795	11.60000	12.00000

Test	F Ratio	DFNum	DFDen	Prob > F
O'Brien[.5]	0.3314	3	16	0.8028
Brown-Forsythe	0.2901	3	16	0.8319
Levene	0.5409	3	16	0.6612
Bartlett	0.1445	3	.	0.9332

Warning: Small sample sizes. Use Caution.

As discussed in the textbook, the Levene test is robust to the assumption of normality, whereas the Bartlett test is extremely sensitive to this assumption. We saw in the previous example that the data appear to have been generated from a process that can be modeled with the normal distribution, so we may use Bartlett's test, which has a p-value of .9332. There is no evidence that the variance of etch rate differs across the levels of the power setting. Further discussion of the tests for equal variances produced by JMP is available from the JMP help documentation.

6. Select **Window > Close All**.

Example 3.5 Analysis of Variance

1. Open Peak-Discharge.jmp.

2. Select **Analyze > Fit Y by X**.

3. Select *Discharge* and click **Y, Response**.

4. Select *Method* and click **X, Factor.**

5. Click **OK**.

6. Click the red triangle next to One-way Analysis of Discharge By Method and select **Unequal Variances**.

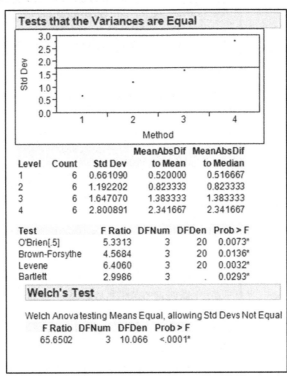

Tests that the Variances are Equal

Level	Count	Std Dev	MeanAbsDif to Mean	MeanAbsDif to Median
1	6	0.661090	0.520000	0.516667
2	6	1.192202	0.823333	0.823333
3	6	1.647070	1.383333	1.383333
4	6	2.800891	2.341667	2.341667

Test	F Ratio	DFNum	DFDen	Prob > F
O'Brien[.5]	5.3313	3	20	0.0073*
Brown-Forsythe	4.5684	3	20	0.0136*
Levene	6.4060	3	20	0.0032*
Bartlett	2.9986	3	.	0.0293*

Welch's Test

Welch Anova testing Means Equal, allowing Std Devs Not Equal

F Ratio	DFNum	DFDen	Prob > F
65.6502	3	10.066	<.0001*

The Levene test rejects the hypothesis of equal variances with a p-value of 0.0032. By default, JMP produces the result of Welch's test, which is a generalization of ANOVA with unequal population variances to factors with more than two levels. Instead, we will apply a variance-stabilizing transformation to the *Discharge* variable.

7. Select **Analyze > Fit Model**.

8. Check **Keep dialog open**. This will enable us to return to the model dialog to make changes to the model.

9. Click **Run**.

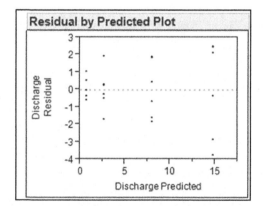

The p-value of the F test is <.0001 indicating that the treatment means are not all equal. However, the Residual by Predicted Plot shows that the assumption of constant variance (homoscedasticity) is violated: the variance of the residuals seems to grow in proportion with the level of discharge (heteroskedasticity). To remedy this, we will take an appropriate transformation, the square root transformation, of *Discharge* and perform an ANOVA on the transformed variable.

10. Return to the **Fit Model** dialog.

11. Select *Discharge* under Pick Role Variables, and then click the red triangle to the right of **Transform**. Select **Sqrt**.

It would also be possible to create an additional column in the data table that contains the transformed values. An advantage of using the transform option, however, is that the predicted values from the Fit Model report are automatically transformed back to the original scale for prediction.

12. Click **Run**.

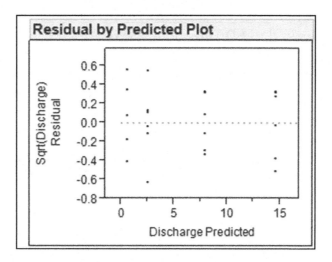

The Residual by Predicted Plot for the ANOVA of the transformed response does not show the same increasing "funnel" of variance that appears in the plot for the analysis of the original response.

13. Select **Window > Close All**.

Example 3.7 Tukey Multiple Comparisons

NOTE: It is possible to perform the multiple comparisons procedures of the next three examples either using the Fit Y by X platform or the Fit Model platform. The next three examples use Fit Y by X, and the example labeled Section 3.8.2 uses Fit Model.

1. Open Etch-Rate.jmp.

2. Select **Analyze > Fit Y by X**.

3. Select *Etch Rate* and select **Y, Response**.

4. Click *Power* and select **X, Factor**.

5. Click **OK**.

6. Click the red arrow next to One-way Analysis of Etch Rate By Power and select **Compare Means > All Pairs, Tukey HSD**.

LSD Threshold Matrix

Abs(Dif)-HSD

	220	200	180	160
220	-33.05	48.55	86.55	122.75
200	48.55	-33.05	4.95	41.15
180	86.55	4.95	-33.05	3.15
160	122.75	41.15	3.15	-33.05

Positive values show pairs of means that are significantly different.

Connecting Letters Report

Level		Mean
220	A	707.00000
200	B	625.40000
180	C	587.40000
160	D	551.20000

Levels not connected by same letter are significantly different.

Ordered Differences Report

Level	-Level	Difference	Std Err Dif	Lower CL	Upper CL	p-Value
220	160	155.8000	11.55335	122.7456	188.8544	<.0001*
220	180	119.6000	11.55335	86.5456	152.6544	<.0001*
220	200	81.6000	11.55335	48.5456	114.6544	<.0001*
200	160	74.2000	11.55335	41.1456	107.2544	<.0001*
200	180	38.0000	11.55335	4.9456	71.0544	0.0216*
180	160	36.2000	11.55335	3.1456	69.2544	0.0294*

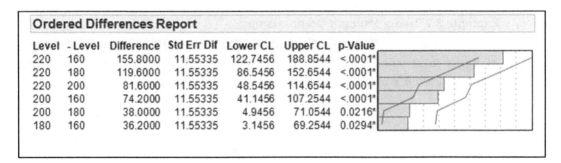

The LSD (least significant difference) Threshold Matrix reports positive values in components that correspond to significantly different pairs of treatment effects. This information is summarized in the Connecting Letters Report. The Ordered Differences Report lists the pairwise comparisons according to magnitude of the difference between the treatment means. The Tukey procedure indicates that all pairs of means are significantly different.

7. Leave the **Fit Y by X** report open for the next two examples.

Example 3.8 Fisher Multiple Comparison

1. Return to the Fit Y by X platform from the previous example.

2. Click the red arrow next to One-way Analysis of Etch Rate By Power and select **Compare Means > Each Pair, Student's t.**

LSD Threshold Matrix

Abs(Dif)-LSD

	220	200	180	160
220	-24.49	57.11	95.11	131.31
200	57.11	-24.49	13.51	49.71
180	95.11	13.51	-24.49	11.71
160	131.31	49.71	11.71	-24.49

Positive values show pairs of means that are significantly different.

Connecting Letters Report

Level					Mean
220	A				707.00000
200		B			625.40000
180			C		587.40000
160				D	551.20000

Levels not connected by same letter are significantly different.

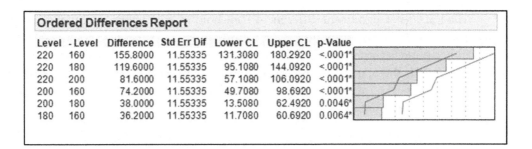

The interpretation of this report is the same as it was for Tukey's test. However, this procedure does not control for the experimentwise error rate. Notice how the confidence intervals for the differences between treatment means are smaller when created using the Fisher Multiple Comparison method than the same intervals that are created when using Tukey's method.

3. Leave the Fit Y by X report open for the next example.

Example 3.9 Dunnett's Multiple Comparison

1. Return to the Fit Y by X platform from the previous example.

2. Click the red arrow next to One-way Analysis of Etch Rate By Power and select **Compare Means > With Control, Dunnett's**.

3. Select 220 for the control group.

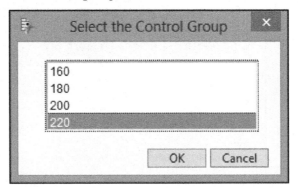

4. Click **OK**.

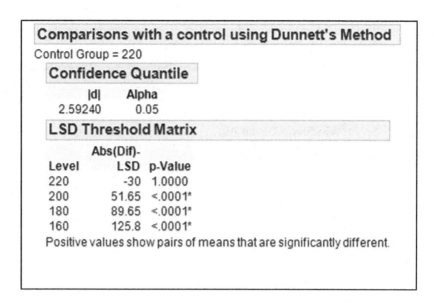

5. The LSD Threshold Matrix indicates that each of the treatment levels is significantly different from the control of 220. Scroll to the top of the Fit Y by X report.

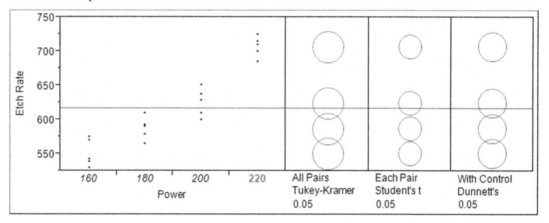

The multiple comparison procedures also produce comparison circles. Group means are significantly different either if their circles do not overlap, or if the angle of their intersection is less than 90 degrees. The comparison circles are meant to provide a quick visual summary of the multiple comparisons procedures. The LSD Threshold Matrices may be consulted to see the details of each comparison.

6. Select **Window > Close All**.

Example 3.10 Power Analysis

1. Select **DOE > Sample Size and Power**.

2. Click **k Sample Means**.

3. Enter 0.01 for **Alpha**.

4. Enter 25 for **Std Dev**.

5. For the first four Prospective Means fields, enter 575, 600, 650, and 675. These means represent the potential means of a response across four different levels for a single factor. When specified along with a value of *Power*, we are asking JMP this question: "If the factor level means are actually equal to these values, we want to reject the null hypothesis (of equality of treatment means) in **Power** x 100 percent of independent replications of this experiment. How many runs must the experiment include to guarantee this?"

6. Enter 0.9 for *Power*.

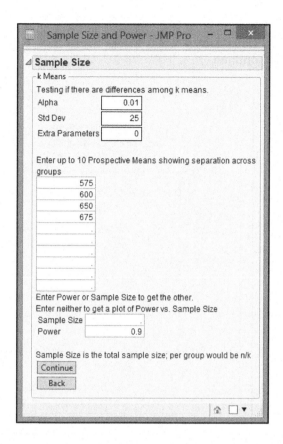

7. Click **Continue**.

8. The resulting total sample size is 15. Dividing this sample size by the four groups and rounding up, we find that 3 replicates of the original 4 experiments (one experiment at each of four factor levels) are necessary to obtain a test with the required power.

9. Clear the values for **Sample Size** and **Power**.

10. Enter 16 for sample size.

11. Click **Continue**.

12. JMP calculates a power of 0.962 for a sample size of 16. This is the power of the test that would result from the experimenter running 3 replicates of a completely randomized experiment on the four factor levels.

13. Delete the values for **Sample Size** and **Power**.

14. Click **Continue**.

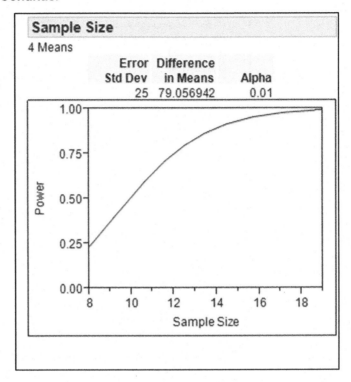

This plot provides a mechanism for determining trade-offs between the sample size and power. The easiest way to view the different values is to select **Tools > Crosshairs**. You can also develop a similar plot for the difference to detect versus sample size or power.

15. Select **Window > Close All**.

Section 3.8.1 Single Factor Experiment

1. Open Chocolate.jmp.

2. Select **Analyze > Fit Y by X**.

3. Select *Antioxidant Capacity* and click **Y, Response**.

4. Select *Factor* and click **X, Factor**.

5. Click **OK**.

6. Click the red triangle next to One-way Analysis of Antioxidant Capacity By Factor and select **Quantiles**.

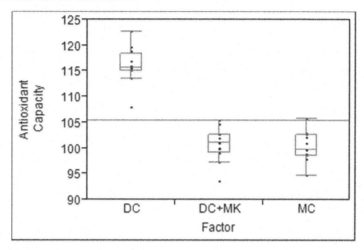

7. Click the red triangle next to One-way Analysis of Antioxidant Capacity By Factor and select **Means/Anova**.

Analysis of Variance

Source	DF	Sum of Squares	Mean Square	F Ratio	Prob > F
Factor	2	1952.6439	976.322	93.5756	<.0001*
Error	33	344.3058	10.434		
C. Total	35	2296.9497			

The p-value of <.0001 for the F test indicates that the treatment means are not all equal.

8. The Fit Y by X platform does not have a built-in functionality for plotting the residuals against the predicted values (though the Fit Model platform does). Click the red triangle next to One-way Analysis of Antioxidant Capacity By Factor and select **Save > Save Residuals**.

9. Click the red triangle next to One-way Analysis of Antioxidant Capacity By Factor and select **Save > Save Predicted**. Alternatively, instead of using the menu twice to select **Save Residuals** and **Save Predicted**, you could hold down the *Alt* key as you click the red triangle next to One-way Analysis of Antioxidant Capacity to select multiple options.

10. Select **Analyze > Fit Y by X.**

11. Select *Antioxidant Capacity centered by Factor* (the residuals) and click **Y, Response**.

12. Select *Antioxidant Capacity mean by Factor* (the predicted values) and click **X, Factor**.

13. Click **OK**.

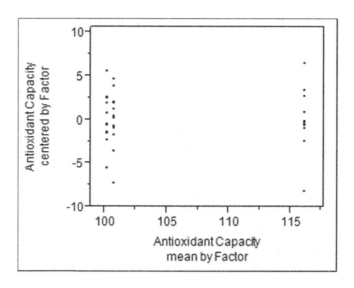

The residual plot does not suggest any violations of the modeling assumptions.

14. Select **Analyze > Distribution**.

15. Select *Antioxidant Capacity centered by Factor* (the residuals) and click **Y, Columns**.

16. Click **OK**.

17. Click the red triangle next to Antioxidant Capacity centered by Factor and select **Normal Quantile Plot**.

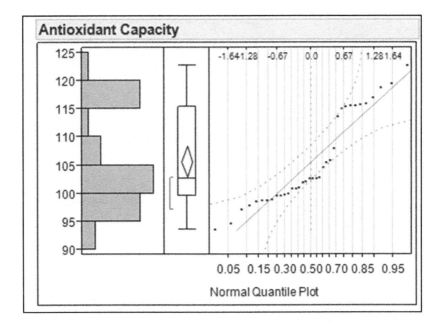

The distribution of the residuals appears to be reasonably close to normal.

18. Return to the original Fit Y by X report (Antioxidant Capacity by Factor).

19. Click the red arrow next to One-way Analysis of Antioxidant Capacity By Factor and select **Compare Means > All Pairs, Tukey HSD**.

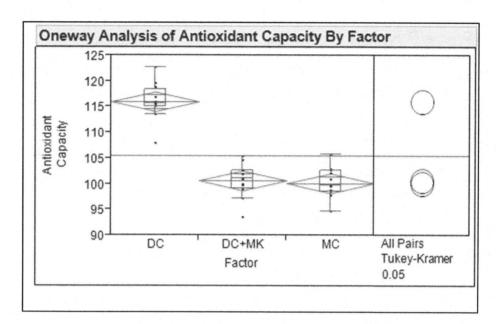

Oneway Analysis of Antioxidant Capacity By Factor

LSD Threshold Matrix

Abs(Dif)-HSD

	DC	DC+MK	MC
DC	-3.236	12.123	12.639
DC+MK	12.123	-3.236	-2.719
MC	12.639	-2.719	-3.236

Positive values show pairs of means that are significantly different.

Connecting Letters Report

Level		Mean
DC	A	116.05833
DC+MK	B	100.70000
MC	B	100.18333

Levels not connected by same letter are significantly different.

Ordered Differences Report

Level	- Level	Difference	Std Err Dif	Lower CL	Upper CL	p-Value
DC	MC	15.87500	1.318681	12.6392	19.11077	<.0001*
DC	DC+MK	15.35833	1.318681	12.1226	18.59410	<.0001*
DC+MK	MC	0.51667	1.318681	-2.7191	3.75244	0.9191

The Tukey test shows that the antioxidant capacity of dark chocolate is significantly greater than both dark chocolate with full-fat milk and milk chocolate. The latter two levels are not significantly different from each other.

20. Select **Window > Close All**.

Section 3.8.2 Application of a Designed Experiment

1. Open Sales-Increase.jmp.

2. From the red triangle next to **Model** in the Table Panel of the data table, click **Run Script**.

3. Click **Run**.

Analysis of Variance

Source	DF	Sum of Squares	Mean Square	F Ratio
Model	2	18.783053	9.39153	35.7722
Error	12	3.150440	0.26254	Prob > F
C. Total	14	21.933493		<.0001*

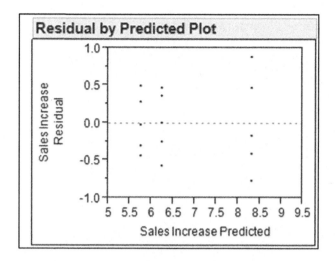

4. Click the red triangle next to Display Design and select **LSMeans Student's t**.

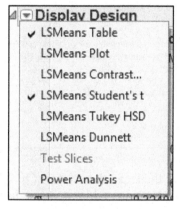

5. Click the red triangle next to LSMeans Differences Student's t and select **Ordered Differences Report.**

LSMeans Differences Student's t

α= 0.050 t= 2.17881

		LSMean[j]		
Mean[i]-Mean[j]	1	1	2	3
Std Err Dif				
Lower CL Dif				
Upper CL Dif				
1		0	-0.506	-2.586
		0	0.32406	0.32406
		0	-1.2121	-3.2921
		0	0.20007	-1.8799
2		0.506	0	-2.08
		0.32406	0	0.32406
		-0.2001	0	-2.7861
		1.21207	0	-1.3739
3		2.586	2.08	0
		0.32406	0.32406	0
		1.87993	1.37393	0
		3.29207	2.78607	0

Level		Least Sq Mean
3	A	8.3180000
2	B	6.2380000
1	B	5.7320000

Levels not connected by same letter are significantly different.

Level	-Level	Difference	Std Err Dif	Lower CL	Upper CL	p-Value	
3	1	2.586000	0.3240597	1.87993	3.292065	<.0001*	
3	2	2.080000	0.3240597	1.37393	2.786065	<.0001*	
2	1	0.506000	0.3240597	-0.20007	1.212065	0.1444	

After noticing that the F test detected that the group means are not all the same, we perform the Fisher LSD procedure and find that the sales increase for design 3 is significantly greater than the increases for designs 1 and 2, and that designs 1 and 2 are not significantly different. Display design 3 shows slightly more variability in sales increases than designs 1 and 2, though the difference may not be significant.

6. Select **Window > Close All**.

Section 3.8.3 Discovering Dispersion Effects

1. Open Smelting.jmp.

2. From the red triangle next to **Model**, click **Run Script**.

3. Select the *s* that appears in the Pick Role Variables section and click the red arrow next to Transform. Select **Log.**

4. Click **Run.**

Analysis of Variance

Source	DF	Sum of Squares	Mean Square	F Ratio
Model	3	6.1660523	2.05535	21.9631
Error	20	1.8716421	0.09358	Prob > F
C. Total	23	8.0376944		<.0001*

5. Click the red arrow next to Ratio Control and select **LSMeans Plot.**

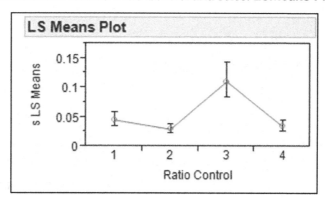

6. Click the red arrow next to Ratio Control and select **LSMeans Tukey HSD.**

```
LSMeans Differences Tukey HSD
Differences are on transformed Y's
a= 0.050   Q= 2.79894

                            Least
Level                      Sq Mean
3     A            0.11082987
1        B         0.04560794
4        B         0.03487751
2        B         0.02993795
Levels not connected by same letter are significantly different.
```

Radio control algorithm 3 produces a significantly higher variation in cell voltage than the other algorithms. We have truncated the output to exclude the table that shows the differences in means and associated confidence intervals.

7. Select **Window > Close All**.

Example 3.11 The Random Effects Model

1. Open Strength-Data.jmp.

2. From the red triangle next to **Model**, click **Run Script**.

3. Select the *Looms* variable that appears in the Construct Model Effects section and click the red triangle next to **Attributes**. Select **Random Effect**.

4. To match the results of the textbook, select **EMS (Traditional)** from the drop-down menu for **Method**. This option tells JMP to use the "method of moments" procedure for fitting the random effects model. However, the recommended restricted maximum likelihood (REML) estimators are often preferred in practice. Maximum likelihood estimation of linear mixed models is covered in Chapter 13.

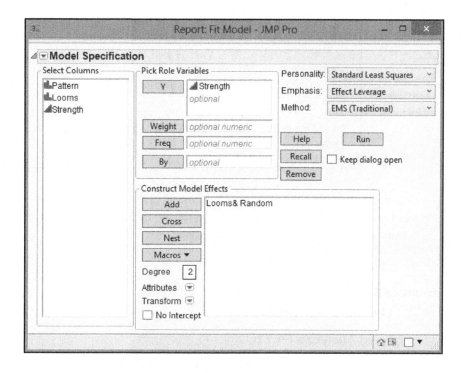

5. Click **Run**.

Analysis of Variance				
Source	DF	Sum of Squares	Mean Square	F Ratio
Model	3	89.18750	29.7292	15.6813
Error	12	22.75000	1.8958	Prob > F
C. Total	15	111.93750		0.0002*

Variance Component Estimates

Component	Var Comp Est	Percent of Total
Looms&Random	6.958333	78.588
Residual	1.895833	21.412
Total	8.854167	100.000

These estimates based on equating Mean Squares to Expected Value.

The F test indicates that the variance component corresponding to *Looms* is significantly greater than 0, meaning that loom-to-loom variation is responsible for a statistically significant portion (78.6%) of the total process variation.

6.   Select **Window > Close All**.

Example 3.12 Nonparametric Analysis

1.   Open Etch-Rate.jmp.

2.   Select **Analyze > Fit Y by X**.

3.   Select *Etch Rate* and select **Y, Response.**

4.   Click *Power* and select **X, Factor**.

5.   Click **OK.**

6.   Click the red arrow next to One-way Analysis of Etch Rate By Power and select **Nonparametric > Wilcoxon Test**. This is another name for the Kruskal-Wallis test, which is also called the Mann-Whitney test when there are only two factor levels present.

Wilcoxon / Kruskal-Wallis Tests (Rank Sums)

Level	Count	Score Sum	Expected Score	Score Mean	(Mean-Mean0)/Std0
160	5	17.000	52.500	3.4000	-3.056
180	5	39.500	52.500	7.9000	-1.091
200	5	63.500	52.500	12.7000	0.917
220	5	90.000	52.500	18.0000	3.231

1-way Test, ChiSquare Approximation

ChiSquare	DF	Prob>ChiSq
16.9070	3	0.0007*

Small sample sizes. Refer to statistical tables for tests, rather than large-sample approximations.

With a p-value of 0.0007, this test rejects the null hypothesis of the equality of the factor level means.

7. Select **Window > Close All**.

Randomized Blocks, Latin Squares, and Related Designs

Chapter 2 introduced the paired t-test for situations where each experimental unit is subjected to both of two treatments under consideration. The difference of the responses from the two treatments is taken within each experimental unit, and the collection of differences is then analyzed with a standard t-test. This process eliminates the unit-to-unit variability from the experimental error. More generally, the unit-to-unit variation may be thought of as coming from a nuisance factor. Nuisance factors are factors that likely affect the response, but are of no interest to the experimenter *per se*.

A nuisance factor that is known and controllable may be managed by blocking on that factor. In a randomized complete block design (RCBD), a replicate of the experiment is run within each block. However, when the blocks are not large enough to contain all of the runs for a single replicate, a randomized incomplete block design is required. An example of an RCBD is presented for measurement of the yield of prosthetic vascular grafts in the presence of a nuisance factor of batches of raw material. An incomplete block design is used to measure four different catalysts when each batch of raw material is only large enough to accommodate three experimental runs.

In situations where two nuisance factors are present, Latin squares may be used to create the experimental design. When p factor levels are being investigated and each of the two blocking factors also have p levels, the rows and columns of a $p \times p$ Latin square correspond to the nuisance factors. Each of the p treatments appears once in each row and once in each column of the square. A 5x5 Latin square design is created using the Custom Design platform. All of the designed experiments presented in this chapter are analyzed using the Fit Model platform.

Example 4.1 A Randomized Complete Block Design

1. Open Vascular-Graft.jmp.

2. From the red triangle next to **Model**, click **Run Script**.

3. Check **Keep dialog open**.

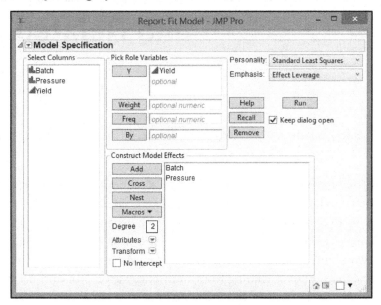

4. Click **Run**.

Summary of Fit	
RSquare	0.771218
RSquare Adj	0.649201
Root Mean Square Error	2.706612
Mean of Response	89.79583
Observations (or Sum Wgts)	24

Analysis of Variance

Source	DF	Sum of Squares	Mean Square	F Ratio
Model	8	370.42333	46.3029	6.3206
Error	15	109.88625	7.3258	Prob > F
C. Total	23	480.30958		0.0011*

Effect Tests

Source	Nparm	DF	Sum of Squares	F Ratio	Prob > F
Batch	5	5	192.25208	5.2487	0.0055*
Pressure	3	3	178.17125	8.1071	0.0019*

With a p-value of 0.0019, the test concludes that, on the average, yield is not equal across extrusion pressures. Notice that JMP returns an F Ratio and a p-value (Prob > F) for *Batch* (blocks), though these should be interpreted with caution because of the restriction on randomization.

It is also possible to treat the blocking variable, *Batch*, as a random effect. This will match the output of Table 4.6 in the textbook.

5. Return to the Fit Model dialog.

6. Select the column *Batch* that appears in the Construct Model Effects section.

7. Click the red triangle to the right of **Attributes** and select **Random Effect**.

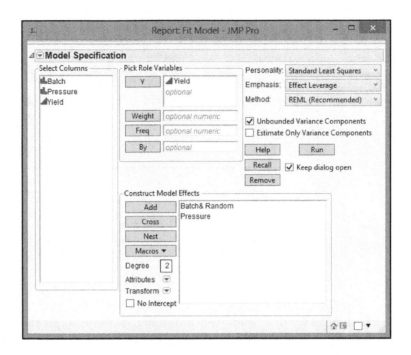

8. Keep the default (REML) **Method**.

9. Click **Run**.

Summary of Fit	
RSquare	0.756688
RSquare Adj	0.720192
Root Mean Square Error	2.706612
Mean of Response	89.79583
Observations (or Sum Wgts)	24

REML Variance Component Estimates

Random Effect	Var Ratio	Var Component	Std Error	95% Lower	95% Upper	Pct of Total
Batch	1.0621666	7.7811667	6.116215	-4.206394	19.768728	51.507
Residual		7.32575	2.6749857	3.9975509	17.547721	48.493
Total		15.106917	6.40202	7.6075741	43.194131	100.000

-2 LogLikelihood = 114.81494867
Note: Total is the sum of the positive variance components.
Total including negative estimates = 15.106917

Covariance Matrix of Variance Component Estimates

Random Effect	Batch	Residual
Batch	37.408085	-1.788887
Residual	-1.788887	7.1555484

The standard errors that are listed for the *Batch* and Residual (error term) effects are the square roots of the diagonal elements of the matrix that appear in the Covariance Matrix of Variance Component Estimates report. This matrix is -1 times the Hessian matrix of the variance components, which is formed by taking the second derivatives of the log-likelihood with respect to the variance components. As discussed in Chapter 13, the Wald confidence intervals based on these values are unreliable when testing whether a variance component is significantly different from 0.

Fixed Effect Tests

Source	Nparm	DF	DFDen	F Ratio	Prob > F
Pressure	3	3	15	8.1071	0.0019*

In this example, the F Ratio for the fixed effect test for *Pressure* is 8.1071 on 3 numerator degrees of freedom (DF) and 15 denominator DF regardless of whether *Batch* is treated as fixed or random. This will not be true in general. For an unbalanced design, the test statistic will be different between the two models, and the denominator DF requires an approximation (e.g., the Satterwaite approximation) in the random effects model.

10. Leave the Vascular-Graft data table open for the next exercise.

Section 4.1.3 Analysis of an Experiment with a Missing Value

The Fit Model platform can perform an exact analysis in the case of missing data. The textbook considers how the analysis would change if the 8700 psi run in the 4th block were missing. An approximate analysis involves imputation of the missing value and a post-hoc reduction of the error degrees of freedom. However, the approximate analysis can be shown to lead to a larger than expected significance level.

1. Return to the Vascular-Graft data table.

2. Delete the *Yield* value for the 8700 psi run in the 4th batch.

	Batch	Pressure	Yield
12	3	9100	85.6
13	4	8500	93.9
14	4	8700	•
15	4	8900	86.2
16	4	9100	87.4
17	5	8500	87.4

3. From the red triangle next to **Model**, click **Run Script**.

4. Click **Run**.

Analysis of Variance

Source	DF	Sum of Squares	Mean Square	F Ratio
Model	8	353.51704	44.1896	6.0834
Error	14	101.69600	7.2640	Prob > F
C. Total	22	455.21304		0.0017*

Effect Tests

Source	Nparm	DF	Sum of Squares	F Ratio	Prob > F
Batch	5	5	189.52200	5.2181	0.0065*
Pressure	3	3	163.39817	7.4981	0.0031*

The approximate analysis that appears in Table 4.8 of the textbook produced an F Ratio of 7.63 and a p-value of 0.0029 for *Pressure*. The least-squares regression and (residual) maximum likelihood estimation routines used by JMP do not require balanced data sets. In general, the exact analysis provided by JMP is a better choice than the approximate analysis presented in the book.

5. Select **Window > Close All**.

Section 4.2 Creating a Latin Square Design in JMP

1. Select **DOE > Custom Design**.

2. Keep the Response Name as *Y*.

3. Select **Add Factor > Categorical > 5 Level**. Change the name from *X1* to *Treatment*.

4. Change the Values for *Treatment* to A, B, C, D, and E.

5. Select **Add Factor > Blocking > Other**. When prompted, specify 5 runs per block.

6. Change the name from *X2* to *Block1*.

7. Select **Add Factor > Blocking > Other**. When prompted, specify 5 runs per block.

8. Change the name from *X3* to *Block2*.

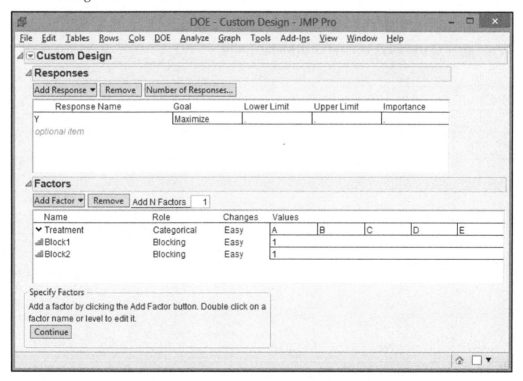

At the moment, only one level appears for each blocking factor. This will be corrected once the **Number of Runs** is specified below.

9. Click **Continue**.

10. Under **Number of Runs**, click the **User Specified option** and enter 25.

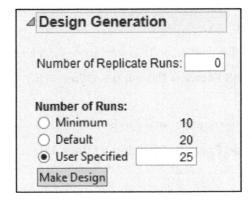

11. Click **Make Design**.

12. Scroll down and click **Make Table**.

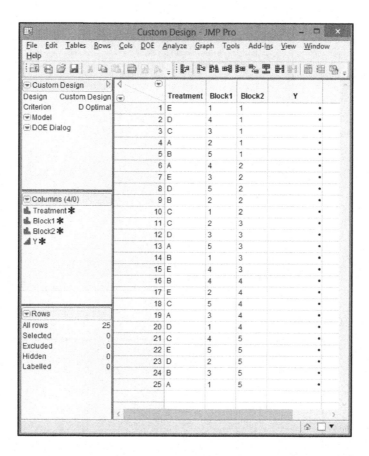

13. To place the design in a square for inspection, select **Tables > Split**.

14. Select *Block1* for **Split By**.

15. Select *Treatment* for **Split Columns**.

16. Select *Block2* for **Group**.

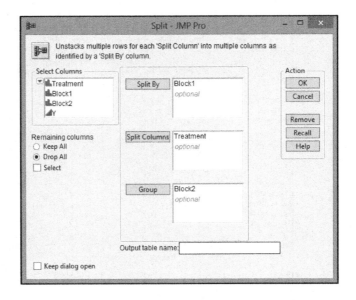

17. Click **OK**.

Block2	1	2	3	4	5
1 1	E	A	C	D	B
2 2	C	B	E	A	D
3 3	B	C	D	E	A
4 4	D	E	A	B	C
5 5	A	D	B	C	E

Notice that each treatment level occurs once in each row and column, as required. Of course, this design presents only one of 161,280 possible 5x5 Latin square configurations (page 162 of the textbook). A different square might be generated by clicking the red triangle next to **Custom Design** after the platform is launched and **Set Random Seed** is selected. Setting the seed to different values should lead to different squares, though there is no guarantee that the optimizer used by the platform will converge to a different square when given a different seed.

18. **Select Window > Close All**.

Example 4.3 Latin Square Design

1. Open Rocket-Propellant.jmp.

2. From the red triangle next to **Model**, click **Run Script**.

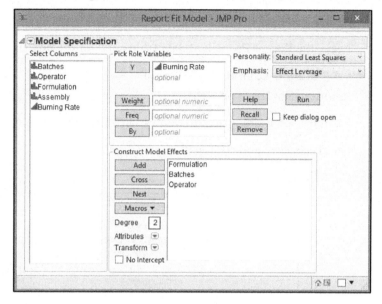

3. Click **Run**.

Analysis of Variance

Source	DF	Sum of Squares	Mean Square	F Ratio
Model	12	548.00000	45.6667	4.2813
Error	12	128.00000	10.6667	**Prob > F**
C. Total	24	676.00000		0.0089*

Effect Tests

Source	Nparm	DF	Sum of Squares	F Ratio	Prob > F
Formulation	4	4	330.00000	7.7344	0.0025*
Batches	4	4	68.00000	1.5938	0.2391
Operator	4	4	150.00000	3.5156	0.0404*

The ANOVA indicates that there is a significant difference in the mean burning rate generated by the different formulations of rocket propellant. With the usual caveat about using caution when testing the significance of blocking factors (due to the restrictions on randomization), the analysis also suggests that there is a sizable operator-to-operator effect, though there seems to be no evidence of batch-to-batch variation.

4. Leave the Rocket-Propellant data table open for the next exercise.

Example 4.4 Graeco-Latin Square Design

1. Return to Rocket-Propellant.jmp.

2. From the red triangle next to **Model**, click **Run Script**.

3. Select *Assembly* and click **Add**.

4. Click **Run**.

Analysis of Variance

Source	DF	Sum of Squares	Mean Square	F Ratio
Model	16	610.00000	38.1250	4.6212
Error	8	66.00000	8.2500	Prob > F
C. Total	24	676.00000		0.0171*

Effect Tests

Source	Nparm	DF	Sum of Squares	F Ratio	Prob > F
Formulation	4	4	330.00000	10.0000	0.0033*
Batches	4	4	68.00000	2.0606	0.1783
Operator	4	4	150.00000	4.5455	0.0329*
Assembly	4	4	62.00000	1.8788	0.2076

Blocking on a factor eliminates the variation due to that factor from the estimated error variance. In this case, the sum of squared errors (SSE) dropped from 128 to 66. However, blocking also reduces the number of degrees of freedom for the F test for the significance of model factors (from 12 to 8 in this application). Blocking on irrelevant factors can reduce the power of the tests for other effects by reducing the error DF (and thus the denominator DF for the F tests) without producing a substantial decrease in the SSE. This is true for blocking in general, not just for Latin or Graeco-Latin squares.

5. Select **Window > Close All**.

Example 4.5 A Balanced Incomplete Block Design

1. Open Catalyst-Experiment.jmp.

2. From the red triangle next to **Model**, click **Run Script**.

3. Check **Keep dialog open**.

4. Click **Run**.

Analysis of Variance

Source	DF	Sum of Squares	Mean Square	F Ratio
Model	6	77.750000	12.9583	19.9359
Error	5	3.250000	0.6500	Prob > F
C. Total	11	81.000000		0.0024*

Effect Tests

Source	Nparm	DF	Sum of Squares	F Ratio	Prob > F
Block	3	3	66.083333	33.8889	0.0010*
Catalyst	3	3	22.750000	11.6667	0.0107*

With a p-value of 0.0107, we conclude that variation in *Catalyst* accounts for a significant portion of the variation in *Time*. The significant *Block* effect indicates that blocking significantly reduced SSE.

5. Click the red triangle next to Catalyst and select **LSMeans Tukey HSD**.

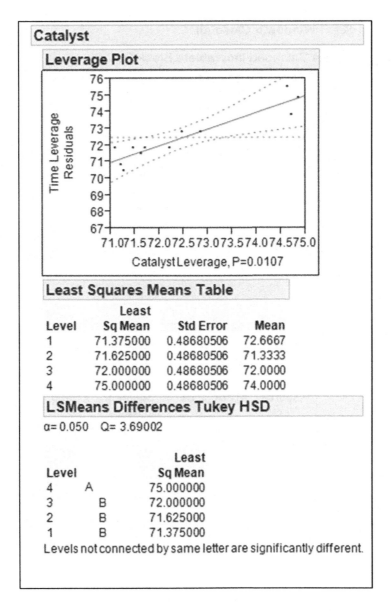

Catalyst 4 appears to be significantly different from the others, which are not significantly different from each other.

6. Click the red triangle next to Response Time and select **Estimates > Expanded Estimates**.

Expanded Estimates

Nominal factors expanded to all levels

Term	Estimate	Std Error	t Ratio	Prob>|t|
Intercept	72.5	0.232737	311.51	<.0001*
Block[1]	0.875	0.427566	2.05	0.0961
Block[2]	3	0.427566	7.02	0.0009*
Block[3]	-3.875	0.427566	-9.06	0.0003*
Block[4]	8.882e-16	0.427566	0.00	1.0000
Catalyst[1]	-1.125	0.427566	-2.63	0.0465*
Catalyst[2]	-0.875	0.427566	-2.05	0.0961
Catalyst[3]	-0.5	0.427566	-1.17	0.2949
Catalyst[4]	2.5	0.427566	5.85	0.0021*

The Estimate column for the levels of *Catalyst* are the intra-block estimates that appear on page 176 of the textbook. We can retrieve the combined estimates (a weighted average of the intra-block and the inter-block estimates) by treating *Block* as a random effect.

7. Return to the **Fit Model** dialog.

8. Select the *Block* column that appears in the Construct Model Effects section.

9. Click the red arrow next to **Attributes** and select **Random Effect**.

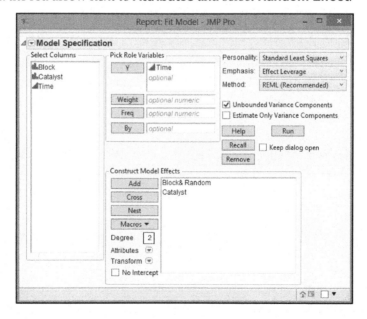

10. Click **Run**.

11. Click the red arrow next to Response Time and select **Estimates > Show Prediction Expression**.

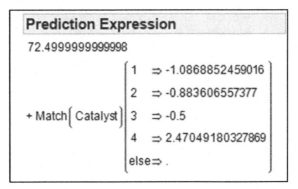

Prediction Expression

72.4999999999998

$+ \text{Match}\left(\text{Catalyst}\right)$
$$\begin{cases} 1 & \Rightarrow -1.0868852459016 \\ 2 & \Rightarrow -0.883606557377 \\ 3 & \Rightarrow -0.5 \\ 4 & \Rightarrow 2.47049180327869 \\ \text{else} \Rightarrow . \end{cases}$$

The values associated with each level of catalyst are the combined estimates of page 176 of the textbook.

12. Select **Window > Close All**.

Introduction to Factorial Designs

This chapter introduces factorial designs for the analysis of multiple experimental factors. In a full factorial design, measurements are made at every possible combination of treatment levels. These designs furnish information about the main effect of each factor as well as interactions between factors. When interactions are significant, the main effects should not be considered in isolation, but in conjunction with the interactions. Interaction plots are useful for interpreting these results.

Factorial designs are often replicated. The replication increases the error degrees of freedom in the ANOVA table and increases the power of the test. However, cost considerations sometimes permit only a single run for each treatment. In this case, there are no error degrees of freedom for the analysis of a full factorial design. If the analyst is willing to assume that the highest order interaction term in the model is insignificant, the mean square value for that interaction term may be used as an estimate of the mean squared error in the F test. If multiple, higher order interactions in the model are insignificant and omitted from the model, the sum-of-squared errors in the F test will be equal to the sum of the sum-of-squared values (from the effect tests) for the insignificant interactions.

Continuous factors may be included by selecting a fixed number of levels from the factor. The factor may then be analyzed either as continuous or categorical. Treating *a* levels from a continuous factor as categorical (with *a* categories) is equivalent to modeling an (*a*-1) degree polynomial of the continuous factor. For example, suppose that factor *x* is measured at 4 levels: 1, 2, 3, and 4. Telling JMP to treat *x* as a nominal (categorical) variable will yield the same model fit as treating *x* as continuous and modeling (intercept + $x + x^2 + x^3$). The behavior of continuous factors may be studied with contour and surface plots when interactions or higher order terms of that factor are included. These ideas are illustrated with an experiment that measures battery life as a function of materials (3 levels) and ambient temperature (3 levels), resulting in a 3^2 factorial design.

In the presence of a nuisance factor, factorial designs may be run in blocks. In a randomized complete block design, each replicate of the experiment occurs within a separate level of the blocking factor. Through the Custom Design platform, JMP provides the ability to design such experiments. The results of the experiment may then be analyzed using the Fit Model platform.

Example 5.1 The Battery Design Experiment

1. Open Battery-Life.jmp.

2. From the red triangle next to **Model**, click **Run Script**.

3. Click **Run**.

Analysis of Variance

Source	DF	Sum of Squares	Mean Square	F Ratio
Model	8	59416.222	7427.03	10.9995
Error	27	18230.750	675.21	Prob > F
C. Total	35	77646.972		<.0001*

Effect Tests

Source	Nparm	DF	Sum of Squares	F Ratio	Prob > F
Temp	2	2	39118.722	28.9677	<.0001*
Material	2	2	10683.722	7.9114	0.0020*
Temp*Material	4	4	9613.778	3.5595	0.0186*

There appears to be a significant interaction between *Temp* and *Material*. An interaction plot will illustrate the nature of the interaction.

4. Click the red triangle next to Response Life and select **Factor Profiling >
    Interaction Plots**.

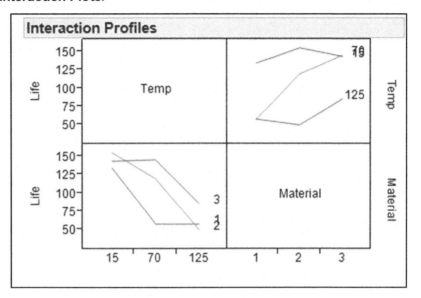

The interaction between the factors is apparent from the non-parallel lines in the
Interaction Profiles plots. All three material types have similar performance at low
temperature. However, as temperature increases and effective battery life decreases,
Material type 3 seems to be the least affected; Material Type 3 has the highest battery life
for Temp 70 and 125.

5. Examine the Residual by Predicted Plot.

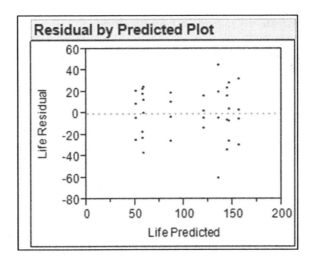

There is slightly greater variability in the residuals for larger predicted battery life times. However, the difference is relatively small. Overall, the plot seems to be satisfactory. There is a potential outlier in the bottom right quadrant of the plot that we will now investigate.

6. Click on the potential outlier. This selects the row corresponding to the point in the data table and highlights the observation in all of the plots that it appears in.

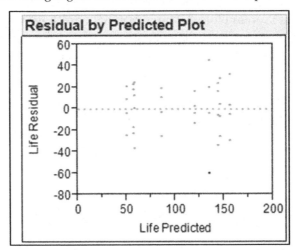

According to the Battery-Life table, the potential outlier is row 14, corresponding to *Temp* of 15 and *Material* 1.

7. Click anywhere in the white space of the Residual by Predicted plot to clear the

selection of the outlier. The other points will change from gray to black.

8. The residuals must be saved to the original data table in order to produce a normal quantile plot and plots of the residuals against the predictors.

9. Click the red triangle next to Response Life and select **Save Columns > Residuals**.

10. Return to the Battery-Life data table and notice that it now contains a column *Residual Life*.

11. Select **Analyze > Distribution**.

12. Select *Residual Life* and click **Y, Columns**.

13. Click **OK**.

14. Click the red triangle next to Residual Life and deselect **Histogram Options > Vertical.**

15. Click the red triangle next to Residual Life and select **Normal Quantile Plot**.

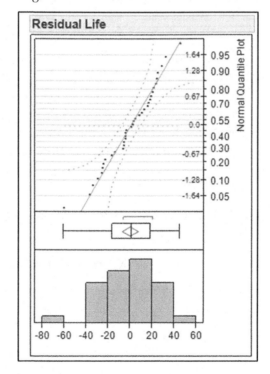

The residuals do not display any departures from normality.

16. Select **Analyze > Fit Y by X**.

17. Select *Residual Life* and click **Y, Response**.

18. Select *Temp* and *Material* and click **X, Factor**.

19. Click **OK**.

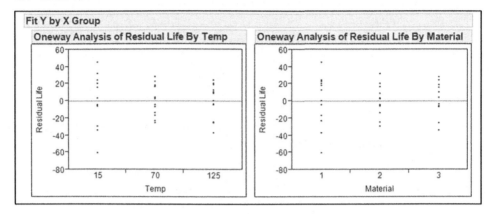

As with the Residual by Predicted plot, there seems to be some evidence of heteroskedasticity (non-constant variance), though it is not severe. The model fit seems to be acceptable. Furthermore, though it is not shown here, if the potential outlier in row 14 were selected, we would see that the smallest residual (bottom left of both plots) is associated with that observation. Without this observation, the variances across the materials and temperatures are not significantly different.

20. Select **Window > Close All**.

Example 5.2 A Two-Factor Experiment with a Single Replicate

1. Open Impurity-Data.jmp.

2. From the red triangle next to **Model**, click **Run Script**.

3. Select the *Temperature*Pressure* interaction and click **Remove**.

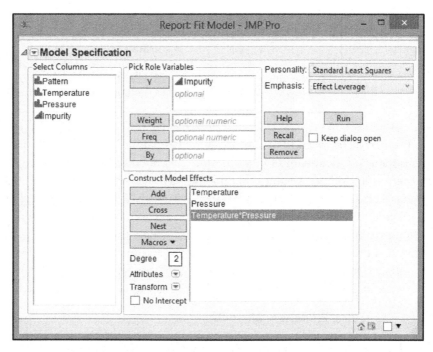

4. Click **Run**.

Summary of Fit

RSquare	0.945848
RSquare Adj	0.905235
Root Mean Square Error	0.5
Mean of Response	2.933333
Observations (or Sum Wgts)	15

Analysis of Variance

Source	DF	Sum of Squares	Mean Square	F Ratio
Model	6	34.933333	5.82222	23.2889
Error	8	2.000000	0.25000	Prob > F
C. Total	14	36.933333		0.0001*

Effect Tests

Source	Nparm	DF	Sum of Squares	F Ratio	Prob > F
Temperature	2	2	23.333333	46.6667	<.0001*
Pressure	4	4	11.600000	11.6000	0.0021*

Both *Temperature* and *Pressure* are significant at the 0.05 level. JMP does not provide an automatic functionality for calculating the nonadditivity test. The test we that have performed simply assumes that the interaction is not significant. Notice that our result has 8 error degrees of freedom, whereas Table 5.11 in the textbook shows only 7 error degrees of freedom. One degree of freedom was used for the nonadditivity test. The information in the table may be used to calculate the test by hand. For this example, the nonadditivity term has a p-value of 0.5674, and we conclude that there is no evidence of interaction between these effects. Afterward, we would reduce the error degrees of freedom from 8 to 7, and recalculate the mean squared errors and F Ratios.

5. Select **Window > Close All**.

Example 5.3 The Soft Drink Bottling Problem

1. Open Soft-Drink.jmp.

2. From the red triangle next to **Model**, click **Run Script**.

3. Click **Run**.

Analysis of Variance

Source	DF	Sum of Squares	Mean Square	F Ratio
Model	11	328.12500	29.8295	42.1123
Error	12	8.50000	0.7083	Prob > F
C. Total	23	336.62500		<.0001*

Effect Tests

Source	Nparm	DF	Sum of Squares	F Ratio	Prob > F
Carbonation	2	2	252.75000	178.4118	<.0001*
Pressure	1	1	45.37500	64.0588	<.0001*
Speed	1	1	22.04167	31.1176	0.0001*
Carbonation*Pressure	2	2	5.25000	3.7059	0.0558
Carbonation*Speed	2	2	0.58333	0.4118	0.6715
Pressure*Speed	1	1	1.04167	1.4706	0.2486
Carbonation*Pressure*Speed	2	2	1.08333	0.7647	0.4869

Speed, Carbonation, and *Pressure* are all significantly associated with the resulting fill volume. The *Carbonation*Pressure* interaction is marginally significant, with a p-value of 0.056.

4. Click the red triangle next to Prediction Profiler and deselect **Desirability Functions.**

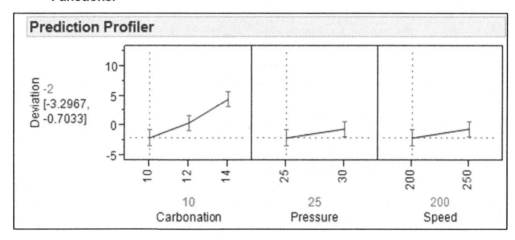

5. Click the red arrow next to Response Deviation and select **Factor Profiling > Interaction Plots**.

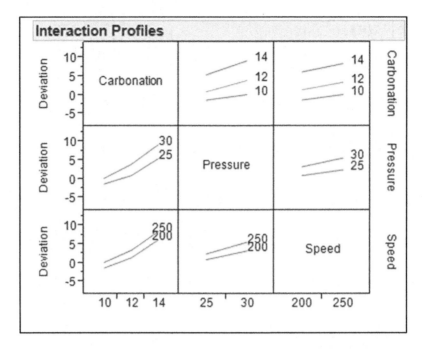

We can see evidence of the marginally significant interaction between *Carbonation* and *Pressure* that was listed in the Effect Tests report. From the top middle square, it is apparent that the *Deviation* is relatively insensitive to *Pressure* when *Carbonation* is set to 10. However, at the higher *Carbonation* level of 14, the *Deviation* increases as *Pressure* increases. If *Pressure* tends to be difficult to control, then using the lower *Carbonation* level will reduce the variation in the observed *Deviation*.

6. Select **Window > Close All**.

Example 5.4 The Battery Design Experiment with a Covariate

1. Open Battery-Life-Covariate.jmp. This example will treat *Temp* as a continuous covariate instead of a categorical design factor. The introduction to this chapter discussed the consequences of treating a covariate as a continuous factor instead of discretizing it into a categorical variable.

2. From the red triangle next to **Model**, click **Run Script**.

3. Select *Life* and click **Y**.

4. Ensure that the **Degree** field lists a value of 2.

5. Select both *Temp* and *Material* and select **Macros > Full Factorial**.

6. Select *Temp* under Select Columns and select **Macros > Polynomial to Degree**.

7. Select *Material* under Select Columns. Under Construct Model Effects, select *Temp*Temp*.

8. Click **Cross**.

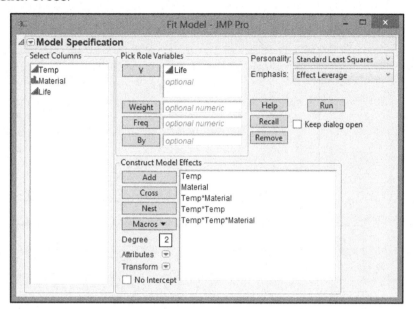

9. Click **Run**.

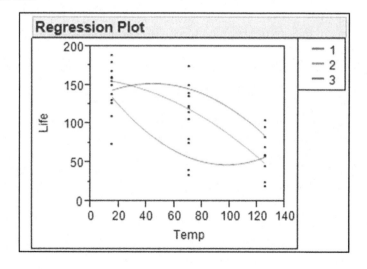

Summary of Fit

RSquare	0.76521
RSquare Adj	0.695642
Root Mean Square Error	25.98486
Mean of Response	105.5278
Observations (or Sum Wgts)	36

Analysis of Variance

Source	DF	Sum of Squares	Mean Square	F Ratio
Model	8	59416.222	7427.03	10.9995
Error	27	18230.750	675.21	Prob > F
C. Total	35	77646.972		<.0001*

Effect Tests

Source	Nparm	DF	Sum of Squares	F Ratio	Prob > F
Temp	1	1	39042.667	57.8227	<.0001*
Material	2	2	16552.667	12.2574	0.0002*
Temp*Material	2	2	2315.083	1.7143	0.1991
Temp*Temp	1	1	76.056	0.1126	0.7398
Temp*Temp*Material	2	2	7298.694	5.4047	0.0106*

The *Temp*Temp*Material* interaction is a significant predictor of battery life. Thus, even though the *Temp*Temp* effect is not significant by itself, that does not imply that a quadratic term in temperature is unimportant; the quadratic effect *depends* on the material type. Therefore, the interaction of this effect with the type of material provides statistically significant information about mean battery life.

10. Click the red triangle next to Response Life and select **Estimates > Expanded Estimates** .

Expanded Estimates

Nominal factors expanded to all levels

Term	Estimate	Std Error	t Ratio	Prob>|t|
Intercept	158.91667	10.09158	15.75	<.0001*
Temp	-0.733333	0.096439	-7.60	<.0001*
Material[1]	-50.33333	10.60827	-4.74	<.0001*
Material[2]	12.166667	10.60827	1.15	0.2615
Material[3]	38.166667	10.60827	3.60	0.0013*
(Temp-70)*Material[1]	0.0310606	0.136385	0.23	0.8216
(Temp-70)*Material[2]	-0.232576	0.136385	-1.71	0.0996
(Temp-70)*Material[3]	0.2015152	0.136385	1.48	0.1511
(Temp-70)*(Temp-70)	-0.001019	0.003037	-0.34	0.7398
(Temp-70)*(Temp-70)*Material[1]	0.0138705	0.004295	3.23	0.0033*
(Temp-70)*(Temp-70)*Material[2]	-0.004642	0.004295	-1.08	0.2894
(Temp-70)*(Temp-70)*Material[3]	-0.009229	0.004295	-2.15	0.0408*

The quadratic temperature effect is positive for *Material* 1 and negative for *Material* 2 and 3. This explains why the Regression Plot for *Material* 1 is concave up while the lines for *Material* 2 and *Material* 3 are concave down.

11. Select **Window > Close All**.

Example 5.5 A 3^2 Factorial Experiment with Two Replicates

1. Open Tool-Life.jmp.

2. From the red triangle next to **Model**, click **Run Script**.

3. Select both *Speed* and *Angle* from the Select Columns area and select **Macros > Response Surface**.

4. Check **Keep Dialog Open**.

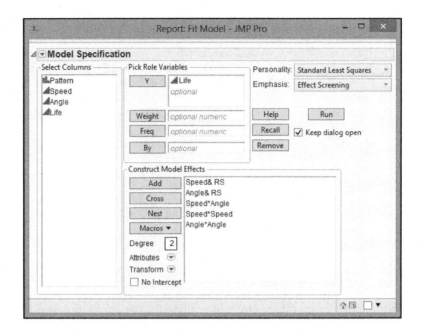

5. Click **Run**.

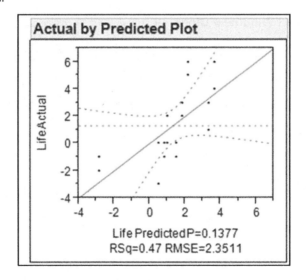

Summary of Fit

RSquare	0.465054
RSquare Adj	0.242159
Root Mean Square Error	2.351123
Mean of Response	1.333333
Observations (or Sum Wgts)	18

Analysis of Variance

Source	DF	Sum of Squares	Mean Square	F Ratio
Model	5	57.66667	11.5333	2.0864
Error	12	66.33333	5.5278	Prob > F
C. Total	17	124.00000		0.1377

Lack Of Fit

Source	DF	Sum of Squares	Mean Square	F Ratio
Lack Of Fit	3	53.333333	17.7778	12.3077
Pure Error	9	13.000000	1.4444	Prob > F
Total Error	12	66.333333		0.0015*

Max R Sq
0.8952

Parameter Estimates

Term	Estimate	Std Error	t Ratio	Prob>|t|
Intercept	-8	5.048683	-1.58	0.1390
Speed	0.0533333	0.027148	1.96	0.0731
Angle	0.1666667	0.135742	1.23	0.2431
(Speed-150)*(Speed-150)	-0.0016	0.001881	-0.85	0.4116
(Speed-150)*(Angle-20)	-0.008	0.00665	-1.20	0.2522
(Angle-20)*(Angle-20)	-0.08	0.047022	-1.70	0.1146

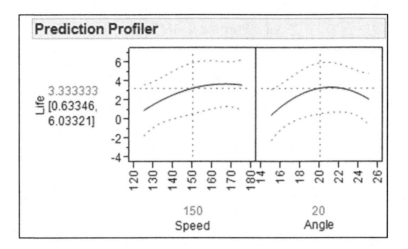

The significant Lack of Fit test with 3 numerator degrees of freedom indicates that at least one of the higher order interaction terms (*Angle*Angle*Speed*, *Angle*Speed*Speed*, and *Angle*Angle*Speed*Speed*) is likely to be important. Furthermore, the apparent scatter in the Actual by Predicted plot along with the RSquare Adj value of 0.24 indicates that the fit of the model is not very good. This is more rigorously demonstrated by the fact that the overall model is not significant, with a p-value of 0.1377 in the Analysis of Variance report. We need to be clear that "fit" and "lack of fit" are two distinct concepts. If the model "fit" is significant, it means that a non-trivial proportion of the process variation has been explained by the model. If the "lack of fit" is significant, it indicates that the functional form assumed by the model is incorrect. We refer to the JMP documentation for a detailed summary of the lack of fit test. It is possible for the model "fit" and "lack of fit" to be simultaneously significant.

6. Return to the **Fit Model** dialog.

7. Select *Speed* under Select Columns. Select *Angle*Angle* under Construct Model Effects.

8. Click **Cross**.

9. Select *Angle* under Select Columns. Select *Speed*Speed* under Construct Model Effects.

10. Click **Cross**.

11. Select *Angle* under Select Columns. Select *Speed*Speed*Angle* under Construct Model Effects.

12. Click **Cross**.

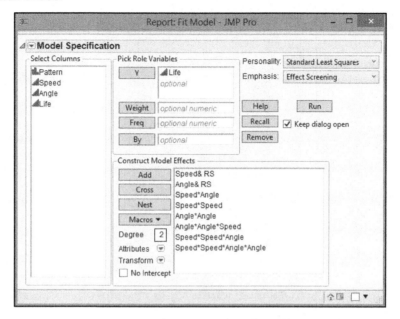

13. Click **Run**.

14. Click the gray triangle next to Effect Tests to expand the report.

15. Click the gray triangles next to Parameter Estimates and Sorted Parameter Estimates in order to hide those reports.

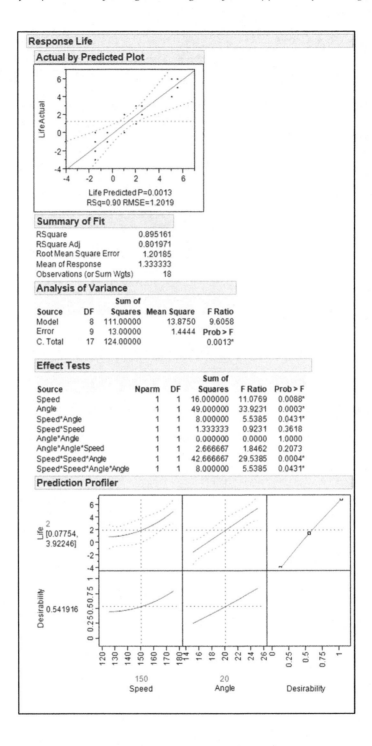

Response Life

Actual by Predicted Plot

Life Predicted P=0.0013
RSq=0.90 RMSE=1.2019

Summary of Fit

RSquare	0.895161
RSquare Adj	0.801971
Root Mean Square Error	1.20185
Mean of Response	1.333333
Observations (or Sum Wgts)	18

Analysis of Variance

Source	DF	Sum of Squares	Mean Square	F Ratio
Model	8	111.00000	13.8750	9.6058
Error	9	13.00000	1.4444	Prob > F
C. Total	17	124.00000		0.0013*

Effect Tests

Source	Nparm	DF	Sum of Squares	F Ratio	Prob > F
Speed	1	1	16.000000	11.0769	0.0088*
Angle	1	1	49.000000	33.9231	0.0003*
Speed*Angle	1	1	8.000000	5.5385	0.0431*
Speed*Speed	1	1	1.333333	0.9231	0.3618
Angle*Angle	1	1	0.000000	0.0000	1.0000
Angle*Angle*Speed	1	1	2.666667	1.8462	0.2073
Speed*Speed*Angle	1	1	42.666667	29.5385	0.0004*
Speed*Speed*Angle*Angle	1	1	8.000000	5.5385	0.0431*

Prediction Profiler

With the addition of the higher order terms, the overall model fit is now significant with a p-value of 0.0013. The RSquare Adj value has increased from 0.24 to 0.80, and the Actual by Predicted plot has improved. The Prediction Profiler report listed above does not match the screenshot in the book. We must take an additional step to match the book.

16. Click the red triangle next to Prediction Profiler and select **Set Desirabilities**.

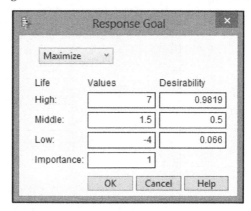

17. Click **OK** to confirm that the objective is to Maximize *Life*.

18. Click the red triangle next to Prediction Profiler and select **Maximize Desirability** .

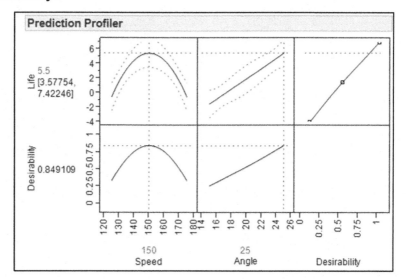

The Prediction Profiler output now matches the screenshot in the book. Setting *Speed* to 150 and *Angle* to 25 will maximize *Life*, with an expected value of 5.5.

19. Click the red triangle next to Response Life and select **Factor Profiling >
    Contour Profiler**.

20. Click the red triangle next to Contour Profile and select **Contour Grid**. In the
    dialog that appears, request the high and low values for the grid contours, along
    with the increment. JMP will suggest reasonable values.

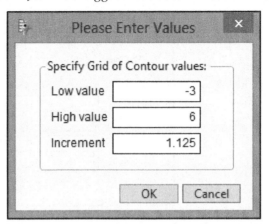

21. Click **OK**.

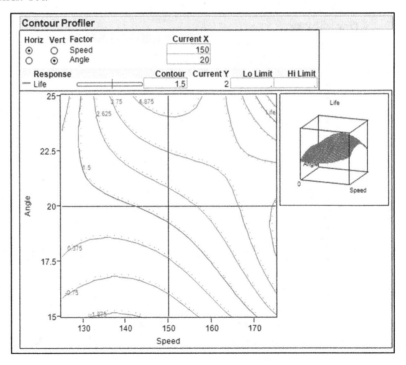

The nonlinear contour plot is a result of the presence of higher order terms of *Speed* and *Angle* (including interaction effects between these factors) in the model.

22. Click the red triangle next to Response Life and select **Factor Profiling > Surface Profiler**.

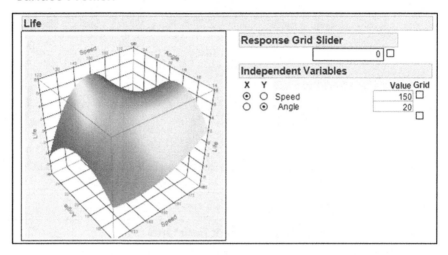

The contour profile seen previously is just the projection of the surface profile onto the two-dimensional *Speed x Angle* design space. In a design with 3 continuous factors, this surface profile may be produced for constant values of the third factor. With more than 3 factors, the surface profile is of limited use for finding optimal factor settings. For these situations, the Desirability Function of the Prediction Profiler provides a better solution.

23. Select **Window > Close All**.

Example 5.6 A Factorial Design with Blocking

1. Open Target-Detection.jmp.

2. From the red triangle next to **Model**, click **Run Script**.

3. Select the *Blocks* column that appears in the Construct Model Effects section.

4. Click the red triangle next to **Attributes** and select **Random Effect.** As described in the JMP documentation, this attribute "creates main effects, two-way interactions, and quadratic terms. Main effects have the response surface attribute, which generates reports that are specific to response surface models."

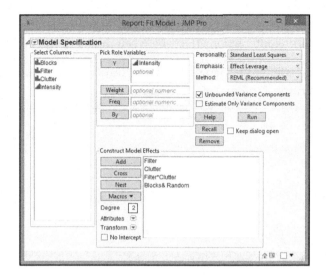

5. Click **Run**.

REML Variance Component Estimates

Random Effect	Var Ratio	Var Component	Std Error	95% Lower	95% Upper	Pct of Total
Blocks	1.8481964	20.494444	18.255128	-15.28495	56.273839	64.890
Residual		11.088889	4.0490897	6.0510389	26.561749	35.110
Total		31.583333	18.552083	12.960706	159.11518	100.000

-2 LogLikelihood = 118.73680261
Note: Total is the sum of the positive variance components.
Total including negative estimates = 31.583333

Fixed Effect Tests

Source	Nparm	DF	DFDen	F Ratio	Prob > F
Filter	1	1	15	96.1924	<.0001*
Clutter	2	2	15	15.1315	0.0003*
Filter*Clutter	2	2	15	3.4757	0.0575

The *Filter*Clutter* interaction is marginally significant. Both *Filter* and *Clutter* are significantly associated with *Intensity*. Operators (*Blocks*) accounted for 64.89% of the residual variation after the mean structure of *Filter* + *Clutter* + *Filter*Clutter* had been removed. Even though the confidence interval for the *Blocks* variance component contains 0, the blocking factor should not be ignored since these confidence intervals tend to be too wide. In this example, the fact that operators account for over half of the process variation provides a sufficient reason to keep the *Blocks* factor. Discussion of the limited utility of Wald confidence intervals for variance components appears in Chapter 13.

6. Select **Window > Close All**.

The 2^k Factorial Design

A popular choice for screening experiments, 2^k factorial designs measure the first-order effects of k factors, as well as interactions between those effects. The "2" indicates that each factor is represented only at a high level and at a low level (or two levels for categorical factors). The restriction to measuring a linear effect of each factor is often reasonable, and provides an efficient design for identifying the significant factors from a large collection of candidates. These designs require 2^k runs.

Analysis of an unreplicated 2^k full factorial design results in a model with no degrees of freedom available for an estimate of error. This chapter demonstrates how normal quantile plots may be used to look for meaningful effects. After you identify inconsequential effects with a normal quantile plot and/or conclude that high order interactions are likely to be insignificant (appealing to the sparsity-of-effects principle), a reduced model may be fit. When analyzed with a reduced model, the 2^k runs provide enough degrees of freedom to obtain an estimate of error, and thus an estimate of the significance of the individual effects. It is also possible to obtain an estimate of error and to test for quadratic curvature by augmenting the 2^k design with center points. This estimate of error is known as "pure error" since it is not a function of insignificant factors.

The 2^k factorial design assumes that the 2^k runs of the experiment are randomized. We distinguish replications of the experiment from duplicate measurements within the experiment. In both situations, multiple observations are made at each treatment combination. When the experiment is replicated, these observations occur in a random order during the execution of the experiment. By contrast, duplicate measurements at each treatment combination are made consecutively after a single run at that combination.

In Example 6.5, an experiment is presented where four wafers are placed in a furnace to study the association between four factors and the resulting oxide thickness on the wafers. An oxide thickness is measured on each of the four wafers during each run, but these four duplicate measurements do not yield the same information that would have been obtained from four replicates of the design. The responses are likely positively correlated. In that case, treating them as independent would produce downward bias in the estimate of error variance, resulting in an inflated number of effects that are incorrectly identified as significant. Short of introducing random effects to account for the within-run correlation, it is appropriate to take an average of the duplicate measurements. The variance of the duplicate measurements may be used to look for factors that affect the within-run variation.

2^k factorial designs are analyzed using linear models (including regression and ANOVA). In JMP, this happens through the Fit Model platform. As discussed in the textbook, it is usually preferable to code the factor levels as high (1) and low (-1) rather than using natural units. This produces an orthogonal design matrix and allows for comparison of the relative importance of factors based on the magnitudes of their estimated effects (the effects are often called scale invariant). JMP automatically codes two-level continuous factors when creating the designs.

Section 6.2 The 2^2 design

We first demonstrate how JMP may be used to create a 2^2 factorial design.

1. Select **DOE > Full Factorial Design**.

2. Double-click **Y** under Response Name and rename the response *Yield*.

3. In the Factors area, select **Continuous > 2 Level**. Set the name to *Conc.* (double click) and change the minimum value to 15 and the maximum to 25.

4. In the Factors area, select **Continuous > 2 Level**. Set the name to *Catalyst*

(double-click) and change the minimum value to 1 and the maximum to 2.

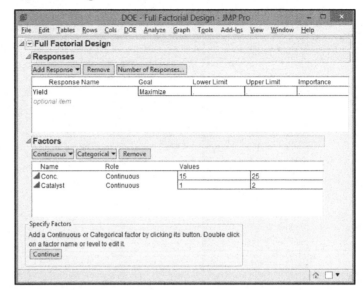

5. Click **Continue**.

6. Set **Number of Replicates** to 2 (signifying 2 replicates of the original set of 4 runs), producing a total of (2+1)*4 = 12 runs.

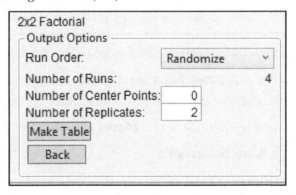

7. Click **Make Table**.

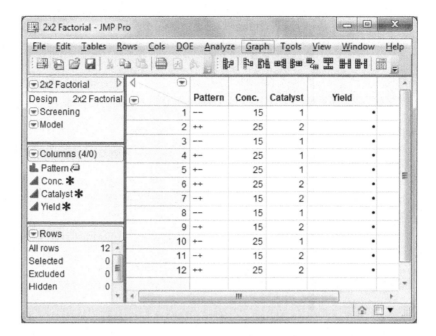

The resulting table is ready to be populated with the results of the experiment. The *Pattern* column is a label column to indicate the high-low settings of the factors. A nice feature of label columns is that a window will pop up when you position the mouse pointer over a point on a plot to indicate the label value of that point. Notice that JMP has randomized the run order for us: we should perform the experiment in the order that the rows appear in this table. The next step is to analyze the data that have been collected.

8. Open Chemical-Process-Yield.jmp. The design, response data, and full factorial model script are contained in this data table.

9. From the red triangle next to **Model**, click **Run Script**.

10. Check **Keep dialog open**.

11. Click **Run**.

Parameter Estimates				
Term	Estimate	Std Error	t Ratio	Prob>|t|
Intercept	27.5	0.571305	48.14	<.0001*
Conc.(15,25)	4.1666667	0.571305	7.29	<.0001*
Catalyst(1,2)	-2.5	0.571305	-4.38	0.0024*
Conc.*Catalyst	0.8333333	0.571305	1.46	0.1828

It is standard practice to remove all insignificant terms from the model and then to re-run the analysis. With a p-value of 0.1828, the *Conc.*Catalyst* interaction does not appear to be significant.

12. Return to the **Fit Model** dialog.

13. Select the *Conc.*Catalyst* effect and click **Remove**.

14. Click **Run**.

Parameter Estimates

Term	Estimate	Std Error	t Ratio	Prob>|t|
Intercept	27.5	0.60604	45.38	<.0001*
Conc.(15,25)	4.1666667	0.60604	6.88	<.0001*
Catalyst(1,2)	-2.5	0.60604	-4.13	0.0026*

15. Scroll down and examine the Residual by Predicted plot. No pattern is discernible.

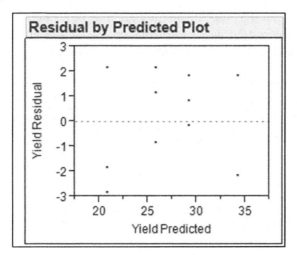

16. To produce surface and contour plots, click the red triangle next to Response Yield and then select **Factor Profiling > Contour Profiler**.

17. Click the red triangle next to Contour Profiler and select **Contour Grid**.

18. Click **OK** to accept the default values for the contour minimum, maximum, and increment.

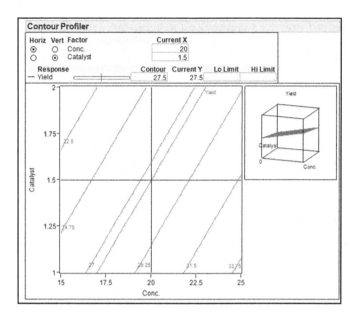

The surface plot that appears on the right may be rotated using the mouse. Our model is linear in *Conc.* and *Catalyst*, producing a plane. Models with higher order terms or with interactions would show curvature (interactions show *bending*) in the surface plot.

19. The Fit Model platform does not automatically produce a Normal Quantile Plot of the residuals. We must first save the residuals as a column in the original data table. Click the red triangle next to Response Yield, and then select **Save Columns > Residuals**.

20. The residuals now appear as a column in the Chemical-Process-Yield data table. To examine the residuals, select **Analyze > Distribution**.

21. Select *Residual Yield* for **Y, Columns**.

22. Click **OK**.

23. Click the red triangle next to Residual Yield and select **Normal Quantile Plot**.

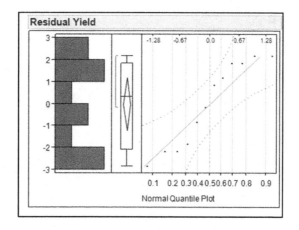

There is no apparent departure from normality in the residuals.

24. Select **Analyze > Fit Y by X**.

25. Select *Residual Yield* and click **Y, Response**.

26. Click *Conc.*, hold down the *Ctrl* key and click *Catalyst*. Click **X, Factor**.

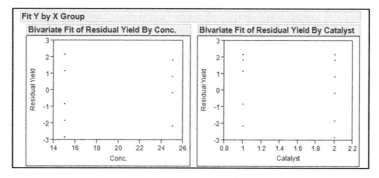

There are no patterns apparent in the plots of residuals against factors, and the variance of the residuals seems to be constant across factor levels.

27. Select **Window > Close All**.

Example 6.1 A 2³ Design

1. Open Nitride-Etch.jmp.

2. From the red triangle next to **Model**, click **Run Script**.

3. Check **Keep dialog open.**

4. Click **Run**.

Parameter Estimates				
Term	Estimate	Std Error	t Ratio	Prob>|t|
Intercept	776.0625	11.86529	65.41	<.0001*
Gap(0.8,1.2)	-50.8125	11.86529	-4.28	0.0027*
Gas Flow(125,200)	3.6875	11.86529	0.31	0.7639
Gap*Gas Flow	-12.4375	11.86529	-1.05	0.3252
Power(275,325)	153.0625	11.86529	12.90	<.0001*
Gap*Power	-76.8125	11.86529	-6.47	0.0002*
Gas Flow*Power	-1.0625	11.86529	-0.09	0.9308
Gap*Gas Flow*Power	2.8125	11.86529	0.24	0.8186

The factors *Gap* and *Power* are significant at the 0.05 level, as is the interaction *Gap*Power*. We will fit a reduced model, removing the insignificant terms. Although it was not an issue in this case, we note it is standard practice to retain an insignificant low-order effect if a higher order of that effect (or an interaction with that effect) is significant; this is commonly referred to as following the Rule of Hierarchy.

5. Return to the **Fit Model** dialog.

6. Delete the insignificant terms from the Construct Model Effects area, leaving only *Gap, Power,* and *Gap*Power*. As a shortcut, notice that first selecting and deleting the *Gas Flow* factor brings up a prompt asking if you would like to remove other effects that contain the selected effect.

7. Click **Run**.

8. Click the red triangle next to Response Etch rate and select **Factor Profiling > Contour Profiler**.

9. Click the red triangle next to Contour Profiler and select **Contour Grid**.

10. Set **Increment** to 100. Click **OK**.

11. Set the value for **Contour** to 900. This displays a line (labeled *Etch Rate*) that enables you to determine which combinations of *Gap* and *Power* result in an etch rate of 900.

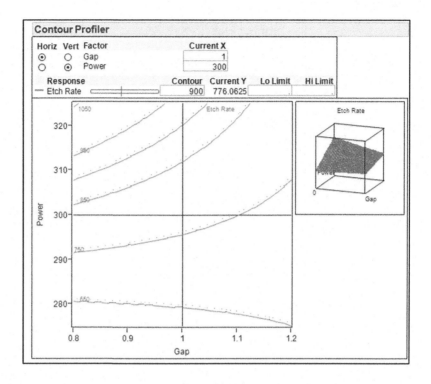

The curvature in the model is a result of the *Gap*Power* interaction. In Step 1, we set the value of Contour to 900, producing the desired contour. **Lo Limit** and **Hi Limit** may also be specified, shading out the out-of-range areas in the design space. Finally, the crosshairs may be moved with the mouse, or controlled with the values that appear under **Current X**.

12. Select **Window > Close All**.

Example 6.2 A Single Replicate of the 2^4 Design

1.   Open Pilot-Plant-Filtration.jmp.

2.   From the red triangle next to **Model**, click **Run Script**.

3.   Check **Keep dialog open**.

4.   Click **Run**.

5.   Click the red triangle next to Response Filtration. Select **Effect Screening > Normal Plot**.

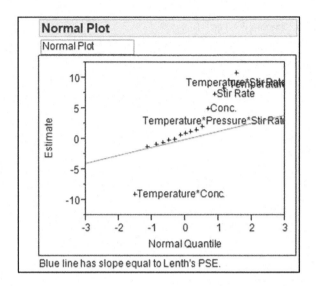

The Normal Plot is a bit cluttered. You can drag the labels to to a different location. JMP will then attach the label to the appropriate point with a line.

Parameter Estimates

Term	Estimate	Std Error	t Ratio	Prob>|t|
Intercept	70.0625	.	.	.
Temperature	10.8125	.	.	.
Pressure	1.5625	.	.	.
Temperature*Pressure	0.0625	.	.	.
Conc.	4.9375	.	.	.
Temperature*Conc.	-9.0625	.	.	.
Pressure*Conc.	1.1875	.	.	.
Temperature*Pressure*Conc.	0.9375	.	.	.
Stir Rate	7.3125	.	.	.
Temperature*Stir Rate	8.3125	.	.	.
Pressure*Stir Rate	-0.1875	.	.	.
Temperature*Pressure*Stir Rate	2.0625	.	.	.
Conc.*Stir Rate	-0.5625	.	.	.
Temperature*Conc.*Stir Rate	-0.8125	.	.	.
Pressure*Conc.*Stir Rate	-1.3125	.	.	.
Temperature*Pressure*Conc.*Stir Rate	0.6875	.	.	.

There are no standard errors associated with the parameter estimates. Since only a single replicate of the design was used, there are no points in the design space with more than one measurement. This means that there are no degrees of freedom for error. Using the Normal Plot, we identify the main effects of *Temperature, Stir Rate,* and *Conc.* as important, as well as the *Temperature*Stir Rate* and *Temperature*Conc.* interactions. The graph also labels the *Temperature*Pressure*Stir Rate* interaction, but it is relatively close to the line and may be considered negligible. In future chapters, we will use the Screening platform to expedite this process.

6. Remove the small effects from the model. Return to the **Fit Model** dialog.

7. Delete all of the effects from the Construct Model Effects area.

8. Select *Temperature* and *Stir Rate* (using the *Ctrl-Click*), and select **Macros > Factorial to Degree**, where **Degree** is set to 2.

9. Repeat Step 8 using *Temperature* and *Conc.*

10. Change the **Emphasis** drop-down selection to Effect Screening if it is not already set.

11. Click **Run**.

12. Scroll down to the Prediction Profiler report.

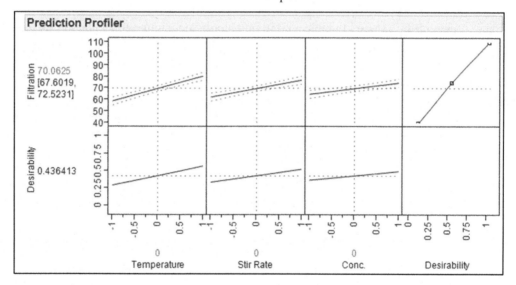

Each of the main effects is positively correlated with *Filtration*. Using this information only, we would maximize *Filtration* by running *Temperature, Stir Rate,* and *Conc.* at their high levels. However, we must also consider the interactions that we found to have large effects.

13. Click the red triangle next to Response Filtration and select **Factor Profiling > Interaction Plots**.

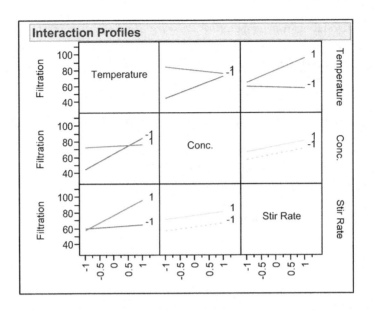

Temperature appears to have a relatively small effect when *Stir Rate* is at its low level, but a much larger effect with *Stir Rate* at its high level. *Filtration* is maximized with *Temperature* = High and *Stir Rate* = High. *Temperature* has a much larger effect when *Conc.* is at its low level than when it is set at its high level. *Filtration* is maximized with *Temperature* = High and *Conc.* = Low. This finding from the interaction profiles contradicts our early conclusion from looking at only the main effects. Notice the interaction effect that was removed from the model, *Stir Rate * Conc.*, has a transparent display in the Interaction Profiler.

14. Return to the Prediction Profiler. Drag the red, dashed vertical lines to change the factor levels. Drag the line for *Temperature* from left to right, and notice how the slopes of the *Filtration* by *Conc.* and *Filtration* by *Stir Rate* plots change as you do so. This is a result of the interaction between *Temperature* and these two factors.

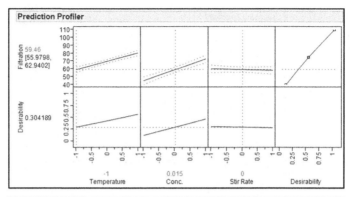

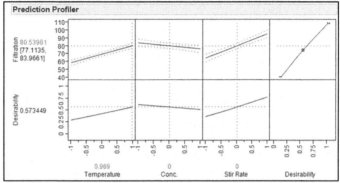

15. Click the red triangle next to Response Filtration and select **Factor Profiling > Contour Profiler**.

16. Select the **Vert** option for *Conc.*, leaving the **Horiz** option selected for *Temperature*.

17. Under **Current X**, set *Stir Rate* to 1.

18. Click the red triangle next to Contour Profiler and select **Contour Grid**.

19. Click **OK** to accept the default minimum, maximum, and increment values for the plot.

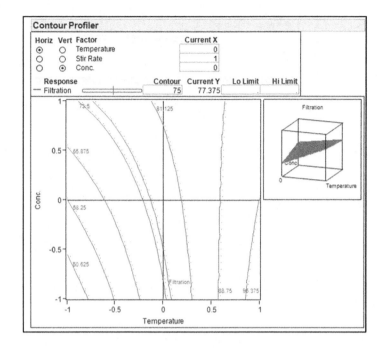

The curvature in this contour plot is a result of the *Temperature*Conc.* interaction. If the **Horiz** factor is changed to *Stir Rate*, the contours will be straight lines since the model does not contain an interaction between *Conc.* and *Stir Rate*.

20. Select **Window > Close All**.

Example 6.3 Data Transformation in a Factorial Design

1.   Open Drilling-Experiment.jmp.

2.   Click the red triangle next to **Model for Example 6.3** in the Left Panel, and select **Run Script.** This will bring up the **Fit Model** dialog.

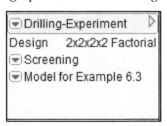

3.   Check **Keep dialog open**.

4.   Click **Run**.

5. Scroll down to the Residual by Predicted Plot. (If it does not appear automatically, it can be produced by clicking the red triangle next to Response Advance Rate and selecting **Row Diagnostics > Plot Residual by Predicted**.)

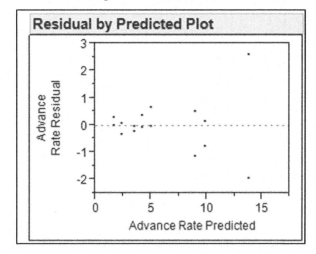

A funnel shape appears in the Residual by Predicted plot, indicating the variance increases as the predicted response increases. This non-constant variance (heteroskedasticity) is a violation of one of the assumptions of the linear model. Although the model is relatively robust for these types of violations, a better model can be fit by transforming the response. This will be discussed further after the normal quantile plot is presented.

6. Produce a normal probability plot: click the red triangle next to Response Advance Rate and select **Save Columns > Residuals**.

7. Select **Analyze > Distribution**.

8. Choose *Residual Advance Rate* for **Y, Columns**.

9. Check **Histograms Only**.

10. Click **OK**.

11. Click the red triangle next to Residual Advance Rate and select **Normal Quantile Plot**.

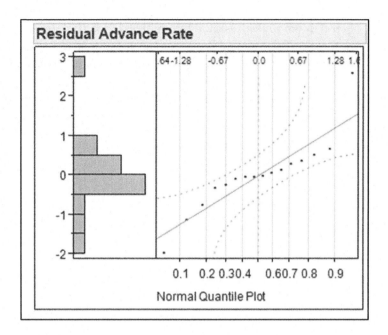

The normal probability plot suggests that the distribution of the residuals is more heavy-tailed than a normal distribution. (The leftmost point on the normal quantile plot is below the line, and the rightmost point is above the line.) In an attempt to address the potential non-normality and heteroskedasticity, We will re-run the analysis using Log(*Advance Rate*) to match the textbook. For guidance selecting the appropriate transformation, see the discussion about the Box-Cox method in Chapter 15.

12. Return to the Fit Model dialog.

13. Select *Advance Rate* under Pick Role Variables.

14. Under Construct Model Effects, click the red triangle next to **Transform** and select **Log**.

15. Delete the interactions *Flow Rate*Speed* and *Flow Rate*Mud* from the model. As mentioned in the text, these interactions no longer appear to be significant after transforming the response variable.

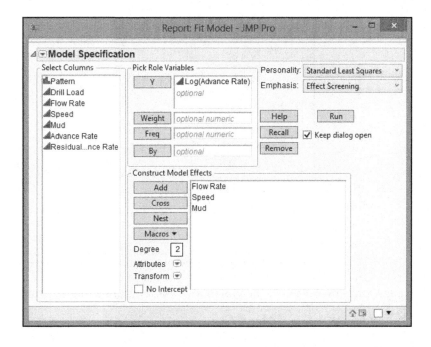

16. Click **Run**.

17. Repeat Steps 5-11 for the results on the transformed data.

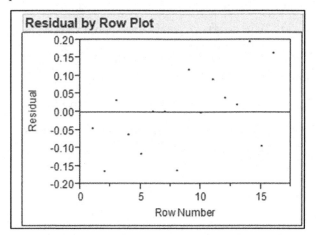

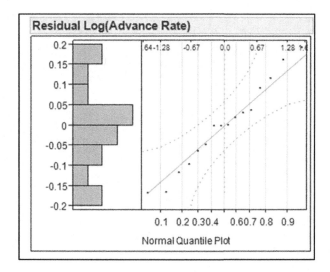

The new plots do not show the violations of assumptions that appeared previously.

18. Select **Window > Close All**.

Example 6.5 Duplicate Measurements on the Response

1. Open Oxide-Thick.jmp. We first create columns for the *Average* and *Variance* of the duplicate measurements made during each run.

2. Double-click the blank column heading to the right of the *Thick4* column.

3. The new column will have a generic name such as *Column 9*. Double-click this column name and change it to *Average*.

4. Right-click *Average* column and select **Formula**.

5. From the Statistical group of functions, select **Mean**.

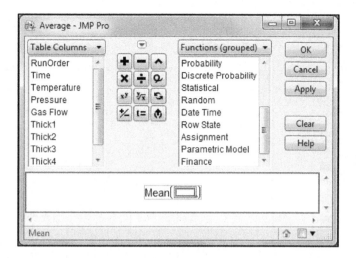

6.   Click *Thick1*, hold down the *Shift* key, and click *Thick4*.

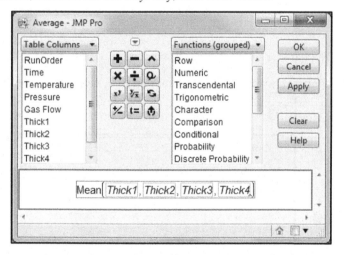

7.   Click **OK**.

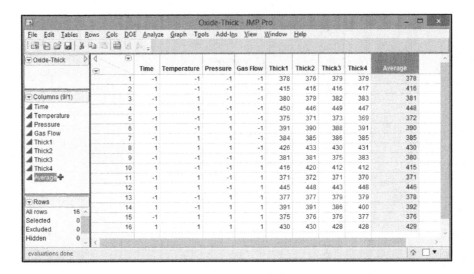

8. Double-click the blank column heading next to *Average*. Name the new column *Variance*.

9. Repeat steps 4-6 for the *Variance* column, but this time use the Std Dev function.

10. Since we want to calculate the variance, select the outer box of Std Dev function. Click 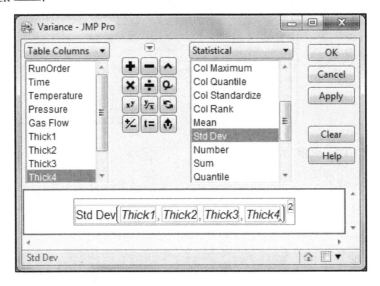.

11. Click **OK**. The data table now matches Table 6.18 from the textbook.

12. Select **Analyze > Fit Model**.

13. For **Y**, select *Average*.

14. Select *Time, Temperature, Pressure,* and *Gas Flow,* and then select **Macros > Full Factorial**.

15. Click **Run**.

16. Click the red triangle next to Response Average. Select **Effect Screening > Normal Plot**.

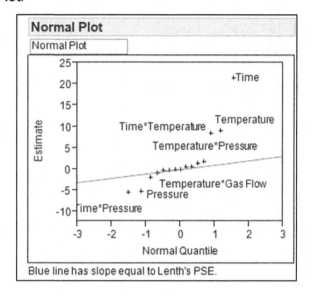

The effects that fall along the line are likely inconsequential. The meaningful effects appear to be *Time, Temperature, Time*Temperature, Pressure,* and *Time*Pressure.* When you are creating this plot, the labels overlap. You can drag the effect labels to reposition them.

17. Click the red triangle next to Response Average and select **Model Dialog**.

18. Delete all but the large effects presented in Step 16.

19. Click **Run**.

20. Click the red triangle next to Response Average and select **Factor Profiling > Contour Profiler**.

21. Click the red triangle next to Contour Profiler and select Contour Grid.

22. Click **OK**.

23. Set the target Contour value to 400.

24. Set Current X for *Pressure* to -1.

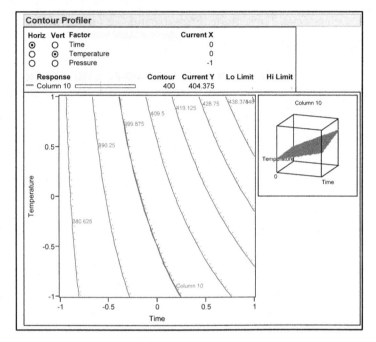

25. Drag the slider for *Pressure* to the right until Current X is 1.

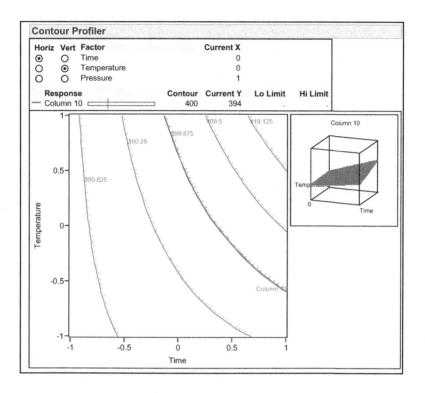

The contour plots show that several combinations of *Time* and *Temperature* will produce an *Average* of 400. However, running the process with *Pressure* at its Low setting allows for shorter run-times, which may be more cost-effective. Later, we will see how JMP offers an option to create desirability functions to optimize multiple responses over several factors.

We next look for factors that affect the within-run variability. Repeating Steps 13-16 with Log(*Variance*) as the response, **Y**, produces a Normal Plot without any reported strong effects. *Time* and *Temperature*Gas Flow* are the largest effects. The log transformation is reasonable here because the distribution of sample variances from an experiment is often right-skewed. The log transformation is also suggested by the Box-Cox method (Chapter 15).

26. Select **Analyze > Fit Model**.

27. Add both *Average* and *Variance* as response variables, **Y**.

28. Select *Variance* under Pick Role Variables and click the red triangle next to **Transform**. Select **Log**.

29. Add *Time, Temperature, Gas Flow,* and *Temperature*Gas Flow* (hold down the *Ctrl* key to highlight both *Temperature* and *Gas Flow* and then click the **Cross** option) under Construct Model Effects.

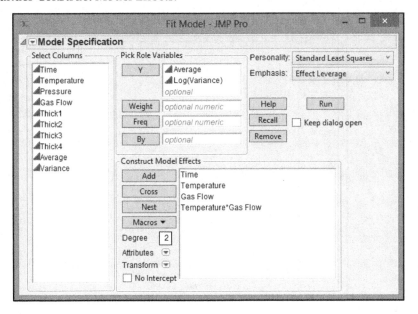

30. Click **Run**.

The experimenters want to keep Average within the range 390 to 410. They also want to keep *Variance* less than 2. JMP provides the ability to construct an overlaid contour plot of both responses (*Average* and *Variance*).

31. Click the red triangle next to Least Squares Fit. Select **Profilers > Contour Profiler**.

32. Set **Current X** for *Gas Flow* to 1.

33. Set the **Hi Limit** for *Variance* to 2.

34. Set the **Lo Limit** and **Hi Limit** for *Average* to 390 and 410, respectively.

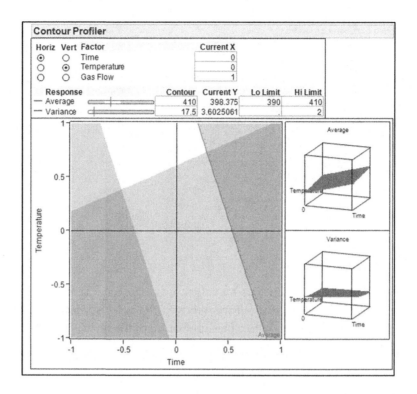

The white region at the top of the plot shows the settings of *Time* and *Temperature* that will yield an in-range average thickness and variance of thickness below 2 when *Gas Flow* is set at the high level. Observe how the white region changes as you drag the setting for *Gas Flow* from 1 to -1.

35. Select **Window > Close All**.

Example 6.6 Credit Card Marketing

1. Open Credit-Card.jmp.

2. From the red triangle next to **Model**, click **Run Script**.

3. Check **Keep dialog open**.

4. Click **Run**.

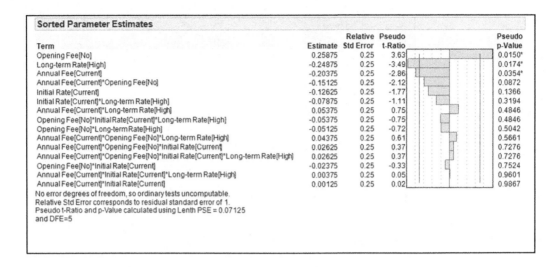

Sorted Parameter Estimates

Term	Estimate	Relative Std Error	Pseudo t-Ratio		Pseudo p-Value
Opening Fee[No]	0.25875	0.25	3.63		0.0150*
Long-term Rate[High]	-0.24875	0.25	-3.49		0.0174*
Annual Fee[Current]	-0.20375	0.25	-2.86		0.0354*
Annual Fee[Current]*Opening Fee[No]	-0.15125	0.25	-2.12		0.0872
Initial Rate[Current]	-0.12625	0.25	-1.77		0.1366
Initial Rate[Current]*Long-term Rate[High]	-0.07875	0.25	-1.11		0.3194
Annual Fee[Current]*Long-term Rate[High]	0.05375	0.25	0.75		0.4846
Opening Fee[No]*Initial Rate[Current]*Long-term Rate[High]	-0.05375	0.25	-0.75		0.4846
Opening Fee[No]*Long-term Rate[High]	-0.05125	0.25	-0.72		0.5042
Annual Fee[Current]*Opening Fee[No]*Long-term Rate[High]	0.04375	0.25	0.61		0.5661
Annual Fee[Current]*Opening Fee[No]*Initial Rate[Current]	0.02625	0.25	0.37		0.7276
Annual Fee[Current]*Opening Fee[No]*Initial Rate[Current]*Long-term Rate[High]	0.02625	0.25	0.37		0.7276
Opening Fee[No]*Initial Rate[Current]	-0.02375	0.25	-0.33		0.7524
Annual Fee[Current]*Initial Rate[Current]*Long-term Rate[High]	0.00375	0.25	0.05		0.9601
Annual Fee[Current]*Initial Rate[Current]	0.00125	0.25	0.02		0.9867

No error degrees of freedom, so ordinary tests uncomputable.
Relative Std Error corresponds to residual standard error of 1.
Pseudo t-Ratio and p-Value calculated using Lenth PSE = 0.07125
and DFE=5

Lenth's Method with simulated *p*-values suggests that all three- and four-way interactions are insignificant. If we remove these effects from the model, they can be used to estimate error for *t*-tests for the significance of the factors and their two-way interactions.

5. Return to the **Fit Model** dialog.

6. Delete the three- and four-way interactions from the Construct Model Effects area.

7. Click **Run**.

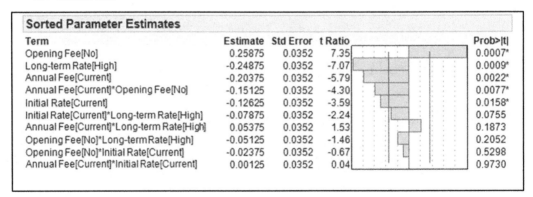

Sorted Parameter Estimates

Term	Estimate	Std Error	t Ratio		Prob>|t|
Opening Fee[No]	0.25875	0.0352	7.35		0.0007*
Long-term Rate[High]	-0.24875	0.0352	-7.07		0.0009*
Annual Fee[Current]	-0.20375	0.0352	-5.79		0.0022*
Annual Fee[Current]*Opening Fee[No]	-0.15125	0.0352	-4.30		0.0077*
Initial Rate[Current]	-0.12625	0.0352	-3.59		0.0158*
Initial Rate[Current]*Long-term Rate[High]	-0.07875	0.0352	-2.24		0.0755
Annual Fee[Current]*Long-term Rate[High]	0.05375	0.0352	1.53		0.1873
Opening Fee[No]*Long-term Rate[High]	-0.05125	0.0352	-1.46		0.2052
Opening Fee[No]*Initial Rate[Current]	-0.02375	0.0352	-0.67		0.5298
Annual Fee[Current]*Initial Rate[Current]	0.00125	0.0352	0.04		0.9730

The four main effects are significant, as well as the *Annual Fee* * *Opening Fee* interaction. We will now further reduce the model to these five effects (four main effects and one interaction) and find the factor settings that maximize the response rate.

8. Return to the **Fit Model** dialog and remove all but the five effects listed in the previous step from the Construct Model Effects area.

9. Click **Run**.

10. Click the red triangle next to Prediction Profiler. Select **Desirability Functions**.

11. Again click the red triangle, and choose **Set Desirabilities**.

12. Select **Maximize** from the drop-down menu if it is not already selected.

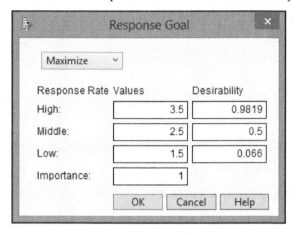

13. Click **OK**.

14. Click the red triangle next to Prediction Profiler, and select **Maximize Desirability**.

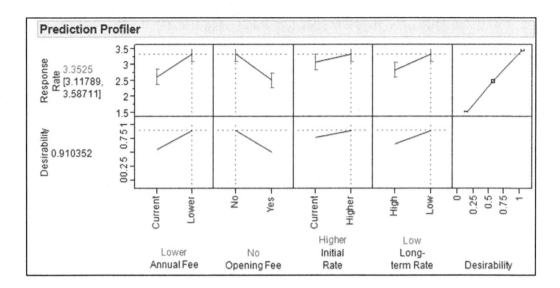

We achieve a maximum *Response Rate* of 3.35 with a lower *Annual Fee,* no *Opening Fee,* a higher *Initial Rate,* and a low *Long-term Rate.*

    15. Select **Window > Close All**.

Example 6.7 A 2⁴ Design with Center Points

    1. Open Filtration-Center.jmp.

    2. From the red triangle next to **Model**, click **Run Script**.

    3. Click **Run**.

Lack Of Fit

Source	DF	Sum of Squares	Mean Square	F Ratio
Lack Of Fit	1	1.512500	1.5125	0.0931
Pure Error	3	48.750000	16.2500	Prob > F
Total Error	4	50.262500		0.7802

Max RSq
0.9916

There is no evidence of second-order curvature in the response over the design space.

Parameter Estimates

Term	Estimate	Std Error	t Ratio	Prob>|t|
Intercept	70.2	0.792642	88.56	<.0001*
Temperature	10.8125	0.886201	12.20	0.0003*
Pressure	1.5625	0.886201	1.76	0.1527
Temperature*Pressure	0.0625	0.886201	0.07	0.9472
Conc.	4.9375	0.886201	5.57	0.0051*
Temperature*Conc.	-9.0625	0.886201	-10.23	0.0005*
Pressure*Conc.	1.1875	0.886201	1.34	0.2513
Temperature*Pressure*Conc.	0.9375	0.886201	1.06	0.3498
Stir Rate	7.3125	0.886201	8.25	0.0012*
Temperature*Stir Rate	8.3125	0.886201	9.38	0.0007*
Pressure*Stir Rate	-0.1875	0.886201	-0.21	0.8428
Temperature*Pressure*Stir Rate	2.0625	0.886201	2.33	0.0805
Conc.*Stir Rate	-0.5625	0.886201	-0.63	0.5601
Temperature*Conc.*Stir Rate	-0.8125	0.886201	-0.92	0.4111
Pressure*Conc.*Stir Rate	-1.3125	0.886201	-1.48	0.2127
Temperature*Pressure*Conc.*Stir Rate	0.6875	0.886201	0.78	0.4812

The significant factors are *Temperature, Stir Rate, Conc., Temperature*Conc.*, and *Temperature*Stir Rate*. We will fit a reduced model with only these factors.

4. Click the red triangle next to Response Filtration and select **Model Dialog**.

5. Delete the insignificant effects. Some analysts prefer to delete a single effect at a time, starting from the highest order interaction and working down (due to the Rule of Hierarchy), performing a backward, stepwise selection. Instead, we will remove all of the terms deemed insignificant by the original model fit.

6. Click **Run**.

Lack Of Fit

Source	DF	Sum of Squares	Mean Square	F Ratio
Lack Of Fit	3	17.13750	5.7125	0.2753
Pure Error	11	228.25000	20.7500	Prob > F
Total Error	14	245.38750		0.8420

Max R Sq
0.9605

Parameter Estimates

Term	Estimate	Std Error	t Ratio	Prob>|t|
Intercept	70.2	0.936154	74.99	<.0001*
Temperature	10.8125	1.046652	10.33	<.0001*
Conc.	4.9375	1.046652	4.72	0.0003*
Stir Rate	7.3125	1.046652	6.99	<.0001*
Temperature*Conc.	-9.0625	1.046652	-8.66	<.0001*
Temperature*Stir Rate	8.3125	1.046652	7.94	<.0001*

7. Select **Window > Close All**.

Blocking and Confounding in the 2^k Factorial Design

This chapter discusses designs that allow 2^k factorial experiments to be run in blocks. When blocks (e.g., batches of raw material) are large enough, a complete replicate of the experiment may be run in each block. In unreplicated designs, at least one of the model effects will be confounded with the blocking factor: it is not possible to differentiate the confounded effect from the blocking effect. Guided by the sparsity-of-effects principle, we are usually most concerned with main effects and two-way interactions, and less concerned with the higher order interactions (three-factor or higher order interactions), which are frequently inconsequential. The standard approach is to confound the blocking effect with the highest order interaction and to assume that it will not be significant using subject-matter knowledge. In general, splitting a 2^k design into 2^p blocks will lead to the confounding of 2^p – 1 effects with the blocking effect.

Confounding may also be present in replicated designs where the block size is too small to contain an entire replicate. This chapter presents an example of a 2^3 design that is split into two blocks, and then run in two replicates. In the first replicate, the highest order interaction $A*B*C$ is confounded with the block effect. However, in the second replicate, the interaction $A*B$ is confounded with the block effect. Consequently, each of the effects may be uniquely estimated. This design is said to be partially confounded: its implementation in JMP is demonstrated through the Screening Design platform.

Example 7.1 A 2^k Replicated Factorial Design with Blocking

1. Open Chemical-Process-Blocked.jmp.

2. From the red triangle next to **Model**, click **Run Script**.

3. Click **Run**.

Analysis of Variance

Source	DF	Sum of Squares	Mean Square	F Ratio
Model	5	298.16667	59.6333	14.4081
Error	6	24.83333	4.1389	Prob > F
C. Total	11	323.00000		0.0027*

Effect Tests

Source	Nparm	DF	Sum of Squares	F Ratio	Prob > F
Conc.(15,25)	1	1	208.33333	50.3356	0.0004*
Catalyst(1,2)	1	1	75.00000	18.1208	0.0053*
Conc.*Catalyst	1	1	8.33333	2.0134	0.2057
Blocks	2	2	6.50000	0.7852	0.4978

As with the example from Section 6.2 where this experiment was run with an unblocked, full factorial design, we see that *Catalyst* and *Conc.* are significant, but their interaction is not. Although the reported p-value seems to indicate that the blocking factor is insignificant, remember that any conclusion based upon this p-value should be viewed with caution because of the lack of randomization across blocks. Heuristically, we may notice that the blocking factor accounts for only 2% of the total process variation (using the percent of Model SS attributed to Block SS).

4. Select **Window > Close All**.

Example 7.2 Blocking and Confounding in an Unreplicated Design

1. Open Filtration-Blocked.jmp.

2. From the red triangle next to **Model**, click **Run Script**.

3. Check **Keep dialog open**.

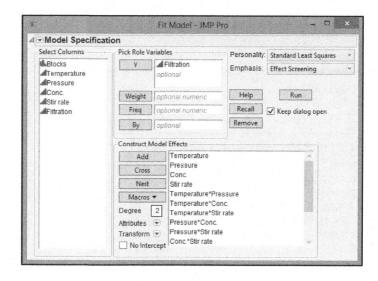

4. Click **Run**.

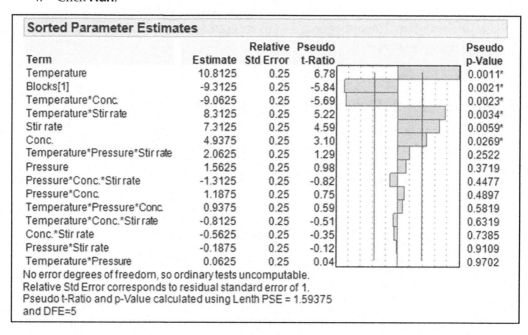

The pseudo p-value from Lenth's test identifies *Temperature, Stir rate, Conc.,
Temperature*Conc.*, and *Temperature* Stir rate* as the large effects. True p-values based on a
t-test are not available for this analysis because there are no degrees of freedom available
for an estimate of error.

5. Return to the Fit Model dialog.

6. Remove all of the effects other than *Blocks, Temperature, Stir rate, Conc., Temperature*Conc.,* and *Temperature* Stir rate* from the Construct Model Effects section.

7. Click **Run**.

Sorted Parameter Estimates					
Term	Estimate	Std Error	t Ratio		Prob>|t|
Temperature	10.8125	1.141279	9.47		<.0001*
Blocks[1]	-9.3125	1.141279	-8.16		<.0001*
Temperature*Conc.	-9.0625	1.141279	-7.94		<.0001*
Temperature*Stir rate	8.3125	1.141279	7.28		<.0001*
Stir rate	7.3125	1.141279	6.41		0.0001*
Conc.	4.9375	1.141279	4.33		0.0019*

Due to the use of an orthogonal design, the parameter estimates are identical between the full and reduced models. However, the reduced model can produce a standard error for the estimates. The orthogonal design also results in identical standard errors for all of the parameter estimates.

8. Select **Window > Close All**.

Example 7.3 A 2^3 Design with Partial Confounding

JMP may be used to generate the design of 7.3, albeit using two steps. The design for replicate I is created, and then combined with the design for replicate II.

1. Select **DOE > Screening Design**.

2. Double-click the response name **Y** and change to *Etch Rate*.

3. Enter 3 in the field next to **Continuous** and click **Add**.

4. Rename the three factors *Gap, Gas Flow,* and *RF*.

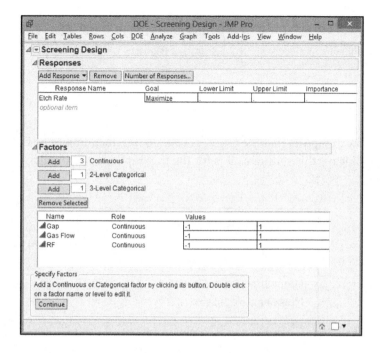

5. Click **Continue**.

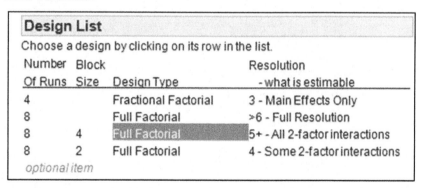

6. Select the design with 8 runs, using 4 runs in 2 blocks.

7. Click **Continue**.

8. Click the gray box next to Change Generating Rules to expand the section.

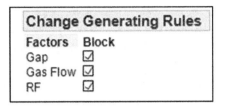

This screen reports that the *Gap*Gas Flow*RF* interaction will be confounded with *Block*. We will leave this setting as it is for the first replicate, and will modify it for the second replicate.

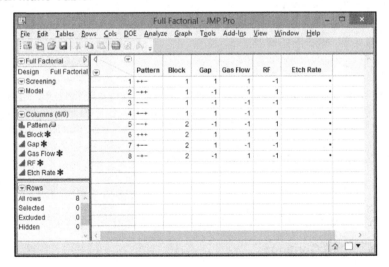

9. Click **Make Table**.

10. Return to the DOE – Screening Design window.

11. Click **Back**.

12. Click **Continue**.

13. Select the design with 8 runs, using 4 runs in 2 blocks.

14. Click **Continue**.

15. Click the gray box next to Change Generating Rules to expand the section.

16. Deselect the box next to *RF*.

17. Click **Apply**.

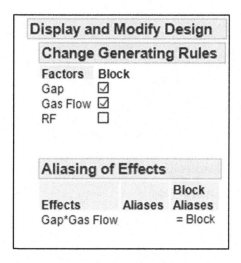

For this replicate, *Gap*Gas Flow* is confounded with *Block*.

18. Click **Make Table**.

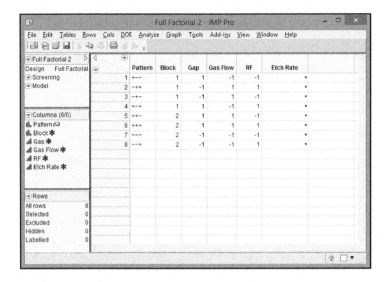

19. Select the *Block* column.

20. Select **Cols > Recode**.

21. Enter 3 and 4 as the **New Value** corresponding to the **Old Value** of 1 and 2, respectively.

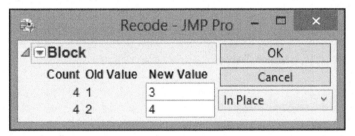

22. Click **OK**.

23. Return to the Full Factorial data table.

24. Select **Tables > Concatenate**.

25. Add *Full Factorial 2* to the dialog box Data Tables to be Concatenated.

26. Enter "Final Design" for Output table name.

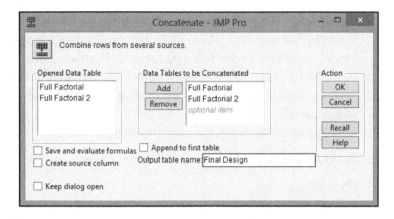

27. Click **OK**.

The completed design now appears in the Final Design data table. Save the data table.

28. Open Plasma-Etch-Partial.jmp. The results from the experiment appear in the *Etch Rate* column of this data table.

29. Select **Analyze > Fit Model**.

30. Click **Run**.

Effect Tests					
Source	**Nparm**	**DF**	**Sum of Squares**	**F Ratio**	**Prob > F**
Gap	1	1	41310.56	16.1941	0.0101*
Gas Flow	1	1	217.56	0.0853	0.7820
RF	1	1	374850.06	146.9446	<.0001*
Block	3	3	5265.69	0.6881	0.5971
Gap*Gas Flow	1	1	3528.00	1.3830	0.2925
Gap*RF	1	1	94402.56	37.0066	0.0017*
Gas Flow*RF	1	1	18.06	0.0071	0.9362
Gap*Gas Flow*RF	1	1	6.13	0.0024	0.9628

The regression analysis indicates that *Gap*, *RF*, and *Gap*RF* are significant at the 0.05 level. This analysis assumes that the four blocks of the two replicates are run sequentially. If there is a large time lag between the two replicates, we may want to decompose the blocking sum of squares into an effect for replicates and a factor for blocks within replicates, as is done within the textbook.

31. Open Plasma-Etch-Partial-2.jmp. This data table contains an additional column, *Replicate*, that indicates which of the two replicates the row came from. The Model script attached to this data table has been modified to include a *Replicate* factor and a *Block[Replicate]* factor, which indicate that *Block* is nested within *Replicate*. In this case, "nested" means that the blocking levels occur uniquely within each *Replicate*. For example, Block 1 in Replicate 1 is different from Block 1 in Replicate 2. Nesting is discussed in detail in Chapter 14.

32. Select **Analyze > Fit Model**.

33. Click **Run**.

Effect Tests

Source	Nparm	DF	Sum of Squares	F Ratio	Prob > F
Gap	1	1	41310.56	16.1941	0.0101*
Gas Flow	1	1	217.56	0.0853	0.7820
RF	1	1	374850.06	146.9446	<.0001*
Block[Replicate]	2	2	1390.63	0.2726	0.7720
Gap*Gas Flow	1	1	3528.00	1.3830	0.2925
Gap*RF	1	1	94402.56	37.0066	0.0017*
Gas Flow*RF	1	1	18.06	0.0071	0.9362
Gap*Gas Flow*RF	1	1	6.13	0.0024	0.9628
Replicate	1	1	3875.06	1.5191	0.2726

The Sum of Squares for *Block[Replicate]* and *Replicate* add up to equal the Sum of Squares for the *Block* effect in the previous analysis. (The numbers in the textbook are slightly different because of a typographical error there.)

34. Select **Window > Close All**.

Two-Level Fractional Factorial Designs

The 2^k factorial designs introduced in Chapter 6 can become very large as the number of factors increase. For example, when $k = 6$, 64 runs are required for the complete 2^k factorial design. These runs furnish the ability to calculate every interaction between the factors, up to the k factor interaction of all effects. However, we are often only concerned with the main factor effects and two-way interactions: only 21 degrees of freedom are associated with these effects. The remaining degrees of freedom are associated with three-factor and higher order interactions.

The two-level, fractional factorial designs introduced in this chapter run only a subset of the complete factorial deigns. The designs attempt to preserve as much information as possible about the main effects and two-factor interactions, while assuming that higher order interactions are negligible. By design, these higher order interactions are confounded with the lower order terms, producing alias chains. The challenge of creating fractional factorial designs is to select subsets of the full factorial design that produce the most desirable alias chains. Given a choice between a design that confounds a main effect with a two-factor interaction and another design that requires the same number of runs but instead confounds a main effect with a high order interaction, we would prefer the latter design since the higher order effect is likely to be insignificant.

Fractional factorial designs are frequently used in screening experiments. In the early stages of process development/improvement, there are often many potential factors of interest, of which only a few may turn out to be important. Further studies may be designed only around the factors found to be significant in the screening experiment. Once a subset of interesting effects are identified, a fold-over design may be run to break alias chains including those effects from the original design. Ultimately, a response surface design (Chapter 11) may be run to optimize the settings of the significant factors.

As discussed in the textbook, three main ideas drive the success of fractional factorial designs:

1. The sparsity-of-effects principle: main effects and low-order interactions often dominate the process.

2. The projection property: once they have been run, factors may be dropped during the analysis of fractional factorial designs, resulting in stronger designs for the remaining factors.

3. Sequential Experimentation: using fold-over designs, additional runs from the full factorial design may be run in a new block after the completion of the original fractional factorial experiment in order to break specific alias chains.

This chapter illustrates how the JMP Screening Design platform may be used to create the fractional factorial designs discussed in the textbook. Once the experiments have been performed, the Screening platform in JMP provides a convenient tool for analyzing fractional factorial designs. Alias chains are automatically detected and reported, and interactions between the factors are produced automatically. According to the JMP documentation, "if your data are all two-level and orthogonal, then all of the statistics in this platform should work well." For blocked designs, the blocking effect should be removed from the response by first fitting a linear model of the response against the (categorical) blocking factor. The Screening platform may then be used with the residuals of this analysis, although it will not recognize that certain effects are confounded with blocks. An illustration of this procedure appears in Example 8.6. As a note of caution, the screening platform does not take into account the model for which the experiment was designed.

In addition to the regular fractional factorial designs, an example of a Plackett-Burman design is presented. The Plackett-Burman designs are non-regular designs in which aliased effects are not completely confounded.

Example 8.1 A Half-Fraction of the 2^4 Design

We first demonstrate how JMP may be used to create a fractional factorial design.

1. Select **DOE > Screening Design**.

2. Double-click the Response Name **Y** and change it to *Filtration*.

3. Enter 4 in the field next to Continuous and click **Add**.

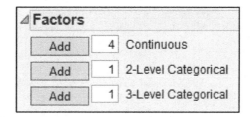

4. Double-click the names of the factors *X1* through *X4* and change them to *Temperature, Pressure, Conc.,* and *Stir Rate,* respectively.

5. Click **Continue**.

6. Select the 8 run Fractional Factorial design (without blocking).

Design List

Choose a design by clicking on its row in the list.

Number Of Runs	Block Size	Design Type	Resolution - what is estimable
8		Fractional Factorial	4 - Some 2-factor interactions
8	4	Fractional Factorial	4 - Some 2-factor interactions
8	2	Fractional Factorial	4 - Some 2-factor interactions
16		Full Factorial	>6 - Full Resolution
16	8	Full Factorial	5+ - All 2-factor interactions
16	4	Full Factorial	4 - Some 2-factor interactions
16	2	Full Factorial	4 - Some 2-factor interactions

optional item

7. Click **Continue**. Expand the Change Generating Rules and Aliasing of Effects reports by clicking the gray triangles.

Display and Modify Design

Change Generating Rules

Factors	Stir Rate
Temperature	☑
Pressure	☑
Conc.	☑

Aliasing of Effects

Effects	Aliases
Temperature*Pressure	= Conc.*Stir Rate
Temperature*Conc.	= Pressure*Stir Rate
Temperature*Stir Rate	= Pressure*Conc.

JMP provides an option to change the defining relation, and displays the aliased effects.

8.   Click **Make Table**.

The design is now loaded into a data table. We will now open the data table that has been populated with the responses.

9.   Open Filtration-Half.jmp.

10.   From the red triangle next to **Model**, click **Run Script**. In the next example, we will demonstrate a similar process using the Screening platform.

Notice that the effects *Pressure*Conc., Pressure*Stir Rate,* and *Conc.*Stir Rate* are not included in the Construct Model Effects area. These effects are confounded with the remaining interactions. If we were to include these effects, JMP would produce the following output.

Singularity Details

Temperature*Stir Rate = Pressure*Conc.
Temperature*Conc. = Pressure*Stir Rate
Temperature*Pressure = Conc.*Stir Rate

11.   Click **Run**.

Sorted Parameter Estimates

Term	Estimate	Relative Std Error	Pseudo t-Ratio		Pseudo p-Value
Temperature	9.5	0.353553	0.77		0.5128
Temperature*Stir Rate	9.5	0.353553	0.77		0.5128
Temperature*Conc.	-9.25	0.353553	-0.75		0.5228
Stir Rate	8.25	0.353553	0.67		0.5649
Conc.	7	0.353553	0.57		0.6213
Pressure	0.75	0.353553	0.06		0.9565
Temperature*Pressure	-0.5	0.353553	-0.04		0.9710

No error degrees of freedom, so ordinary tests uncomputable.
Relative Std Error corresponds to residual standard error of 1.
Pseudo t-Ratio and p-Value calculated using Lenth PSE = 12.375
and DFE=2.3333

The *t*-statistics cannot be calculated for these effects because there are no degrees of freedom available for an estimate of error. The Pseudo p-values result from Lenth's method. Judging from the relative magnitudes of the estimates, it appears that the main effects *Temperature, Conc.,* and *Stir Rate* are large. In this case, the principle of strong heredity would suggest that the *Temperature*Conc.+Pressure*Stir Rate* and the *Temperature*Stir Rate+Pressure*Conc.* alias chains are large because of the *Temperature*Conc.* and *Temperature*Stir Rate* interactions.

12. Click the red triangle next to Response Filtration and select **Factor Profiling > Cube Plots**.

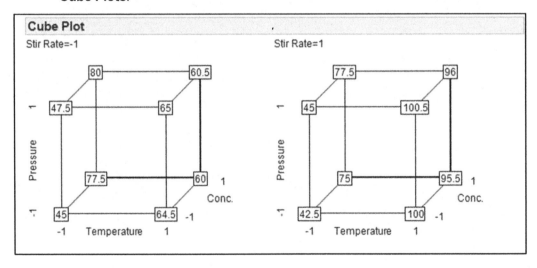

13. Click the red triangle next to Response Filtration and select **Model Dialog**.

14. Delete *Pressure* from the Construct Model Effects. When prompted, select **Yes** for the question "Remove other effects containing selected effect in the model?"

15. Click **Run**.

Parameter Estimates				
Term	Estimate	Std Error	t Ratio	Prob>|t|
Intercept	70.75	0.637377	111.00	<.0001*
Temperature	9.5	0.637377	14.90	0.0045*
Conc.	7	0.637377	10.98	0.0082*
Stir Rate	8.25	0.637377	12.94	0.0059*
Temperature*Conc.	-9.25	0.637377	-14.51	0.0047*
Temperature*Stir Rate	9.5	0.637377	14.90	0.0045*

16. Select **Window > Close All**.

Example 8.2 A 2^{5-1} Design Used for Process Improvement

1. Open Integrated-Circuit.jmp.

2. Select **Analyze > Modeling > Screening**. This platform identifies aliased factors in orthogonal designs. We refer to the JMP documentation for further discussion of the Screening platform.

3. Select *Yield* for **Y**.

4. Select *Aperture, Exposure, Develop, Mask,* and *Etch*. Click **X**.

5. Click **OK**.

Contrasts

Term	Contrast		Lenth t-Ratio	Individual p-Value	Simultaneous p-Value
Exposure	16.9375		36.13	<.0001*	<.0001*
Aperture	5.5625		11.87	0.0002*	0.0005*
Develop	5.4375		11.60	0.0002*	0.0005*
Mask	-0.4375		-0.93	0.3264	0.9983
Etch	0.3125		0.67	0.5315	1.0000
Exposure*Aperture	3.4375		7.33	0.0010*	0.0064*
Exposure*Develop	0.3125		0.67	0.5315	1.0000
Aperture*Develop	0.1875		0.40	0.7146	1.0000
Exposure*Mask	-0.0625		-0.13	0.8984	1.0000
Aperture*Mask	0.5625		1.20	0.2172	0.9576
Develop*Mask	0.4375		0.93	0.3264	0.9983
Exposure*Etch	-0.0625		-0.13	0.8984	1.0000
Aperture*Etch	0.5625		1.20	0.2172	0.9576
Develop*Etch	0.1875		0.40	0.7146	1.0000
Mask*Etch	-0.6875		-1.47	0.1472	0.8255

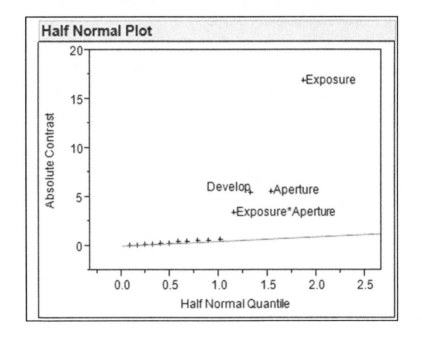

The half normal plot indicates that the effects for *Aperture, Exposure, Develop,* and *Exposure*Aperture* are large. Relying on the sparsity-of-effects principle, we assume that the higher order interactions (three-factor interactions and higher) are insignificant. We will next fit a reduced model using only the four terms that are identified by the half normal plot.

6. Scroll to the bottom of the Screening platform and click **Make Model**.

7. Click **Run**.

8. Click the gray triangle next to Effect Tests to expand the report.

Effect Tests

Source	Nparm	DF	Sum of Squares	F Ratio	Prob > F
Aperture	1	1	495.0625	193.1951	<.0001*
Exposure	1	1	4590.0625	1791.244	<.0001*
Develop	1	1	473.0625	184.6098	<.0001*
Aperture*Exposure	1	1	189.0625	73.7805	<.0001*

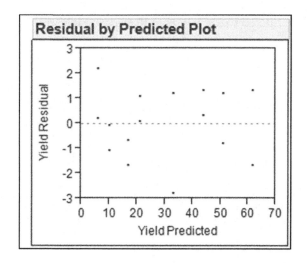

All of the effects in the reduced model are significant, with p-values <0.0001. From the Residual by Predicted Plot, there does not appear to be a violation of the assumption of equal variance.

9. Click the red triangle next to Response Yield and select **Factor Profiling > Interaction Plots**.

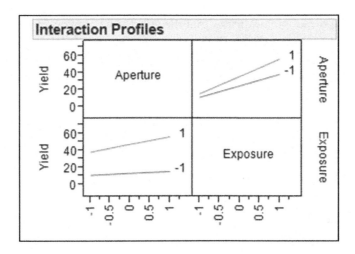

The *Aperture*Exposure* interaction is apparent in the different slopes that appear in the Interaction Profiles. *Yield* is relatively insensitive to *Aperture* when *Exposure* is set to its low level. However, *Yield* shows a stronger increase with *Aperture* when *Exposure* is set to its high level.

10. Click the red triangle next to Response Yield and select **Factor Profiling > Cube Plots**.

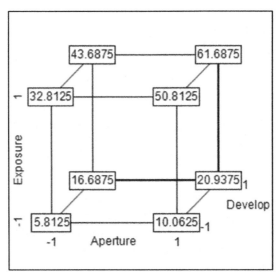

The factors *Aperture, Exposure,* and *Develop* have large, positive effects. The *Aperture*Exposure* interaction plot shows *Yield* is maximized when both *Aperture* and *Exposure* are at their high levels. Setting all three factors to their high level maximizes the mean *Yield* in the cube plot.

11. Click the red triangle next to Response Yield and select **Save Columns > Residuals**.

12. Return to the Integrated-Circuit data table and notice the new column, *Residual Yield*.

13. Choose **Analyze > Distribution**.

14. Select *Residual Yield* for **Y, Columns**.

15. Click **OK**.

16. Click the red triangle next to Residual Yield and deselect **Histogram Options > Vertical**.

17. Click the red triangle next to Residual Yield and select **Normal Quantile Plot**.

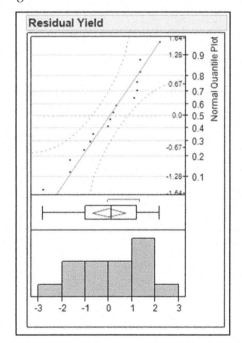

There does not appear to be a violation of the assumption of normality.

18. Select **Window > Close All**.

Example 8.3 A 2^{4-1} Design with the Alternate Fraction

1. Open Filtration-Half-Alt.jmp. This design matrix may be obtained from the one in Filtration-Half.jmp via the DOE > Augment Design platform using the **Foldover** option on *Stir Rate*. The Augment Design platform is illustrated in Example 8.7.

2. Click the red triangle next to the **Screening** script in the left panel and select **Run Script**. This script is added to the data table by the Screening Design platform. For imported data, select **Analyze > Modeling > Screening**.

Contrasts

Term	Contrast		Lenth t-Ratio	Individual p-Value	Simultaneous p-Value	Aliases
Temperature	12.1250		1.27	0.1815	0.7086	
Stir Rate	6.3750		0.67	0.5002	1.0000	
Conc.	2.8750		0.30	0.8011	1.0000	
Pressure	2.3750		0.25	0.8371	1.0000	
Temperature*Stir Rate	7.1250		0.75	0.3989	0.9970	Conc.*Pressure
Temperature*Conc.	-8.8750		-0.93	0.2990	0.9396	Stir Rate*Pressure
Stir Rate*Conc.	-0.6250		-0.07	0.9569	1.0000	Temperature*Pressure

The *Temperature*Stir Rate* and *Temperature*Conc.* effects are relatively large compared to the other effects, but not to the noise in the process.

3. Click the *Temperature* Term in the Contrasts report. Hold down the *Shift* key and click the *Temperature*Conc.* term to select all of the terms between these two.

4. Scroll to the bottom of the Screening platform and click **Make Model**.

5. Click **Run**.

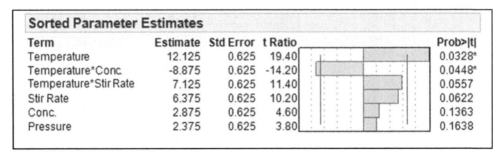

Sorted Parameter Estimates

Term	Estimate	Std Error	t Ratio		Prob>|t|
Temperature	12.125	0.625	19.40		0.0328*
Temperature*Conc.	-8.875	0.625	-14.20		0.0448*
Temperature*Stir Rate	7.125	0.625	11.40		0.0557
Stir Rate	6.375	0.625	10.20		0.0622
Conc.	2.875	0.625	4.60		0.1363
Pressure	2.375	0.625	3.80		0.1638

Now that an estimate of error is available, we see that *Temperature* and *Temperature*Conc.* are significant at the 0.05 level. Even though Lenth's method did not flag any important effects, we were able to find significant factors when looking at a reduced model.

6. Select **Window > Close All**.

Example 8.4 A 2^{6-2} Design

1. Open Injection-Molding.jmp.

2. Click the red triangle next to the **Screening** script in the table panel and select **Run Script**. This script is added to the data table by the Screening Design platform. For imported data, select **Analyze > Modeling > Screening**.

Contrasts						
Term	Contrast		Lenth t-Ratio	Individual p-Value	Simultaneous p-Value	Aliases
Screw	17.8125		38.00	<.0001*	<.0001*	Tempera
Temperature	6.9375		14.80	<.0001*	0.0003*	Screw*H
Cycle Time	0.6875		1.47	0.1441	0.8277	Screw*H
Hold Time	-0.4375		-0.93	0.3266	0.9975	Screw*Te
Gate	0.1875		0.40	0.7157	1.0000	Screw*Te
Pressure	0.1875		0.40	0.7157	1.0000	Screw*Cy
Screw*Temperature	5.9375		12.67	<.0001*	0.0004*	Hold Tim
Screw*Cycle Time	-0.0625		-0.13	0.9067	1.0000	Hold Tim
Temperature*Cycle Time	-2.6875		-5.73	0.0018*	0.0175*	Gate*Pre
Screw*Hold Time	-0.9375		-2.00	0.0621	0.4752	Tempera
Temperature*Hold Time	-0.8125		-1.73	0.0947	0.6513	Screw*G
Cycle Time*Hold Time	-0.0625		-0.13	0.9067	1.0000	Screw*Pr
Cycle Time*Gate	0.3125		0.67	0.5315	1.0000	Tempera
Screw*Temperature*Cycle Time	0.0625		0.13	0.9067	1.0000	Cycle Tim
Temperature*Cycle Time*Hold Time	-2.4375		-5.20	0.0024*	0.0227*	Screw*C

The larges effects are *Temperature, Screw,* and *Temperature*Screw*. We will now fit a reduced model using only these effects. One option is to customize the selected terms in the Contrasts report by clicking terms while holding down the *Ctrl* key and then choosing the **Make Model** option at the bottom of the Screening platform. Instead, we will work within the **Fit Model** dialog.

3. From the red triangle next to **Model**, click **Run Script**.

4. Remove all of the terms from the Construct Model Effects area.

5. Select *Temperature* and *Screw* and select **Macros > Full Factorial**.

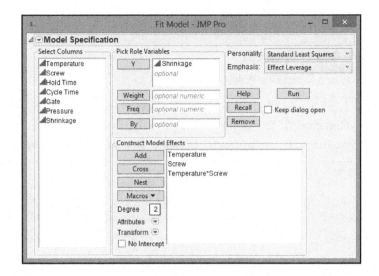

6.   Click **Run**.

Parameter Estimates

Term	Estimate	Std Error	t Ratio	Prob>|t|
Intercept	27.3125	1.138232	24.00	<.0001*
Temperature	6.9375	1.138232	6.09	<.0001*
Screw	17.8125	1.138232	15.65	<.0001*
Temperature*Screw	5.9375	1.138232	5.22	0.0002*

7.   Click the red triangle next to Response Shrinkage and select **Estimates > Show Prediction Expression**.

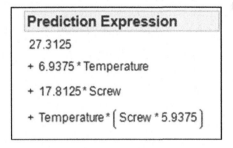

Even though we may use this formula to obtain predictions for values of *Temperature* or *Screw* outside of the range of [-1,1], the prediction variance expands significantly as you leave the design space. Always use extrapolation with caution.

8. Click the red triangle next to Response Shrinkage and select **Factor Profiling > Interaction Plots**.

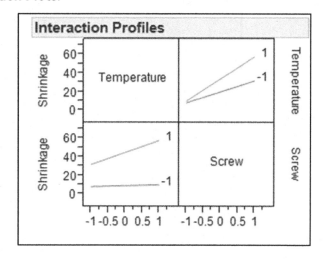

The interaction plot shows that the relationship between *Shrinkage* and *Temperature* is weak when screw speed is at its low level, but grows stronger when the screw speed increases.

9. Click the red triangle next to Response Shrinkage and select **Save Columns > Residuals**.

10. Choose **Analyze > Distribution**.

11. Select *Residual Shrinkage* for **Y, Columns**.

12. Click **OK**.

13. Click the red triangle next to Residual Shrinkage and deselect **Histogram Options > Vertical**.

14. Click the red triangle next to Residual Shrinkage and select **Normal Quantile Plot**.

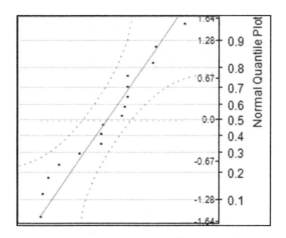

There are no abnormalities in the normal quantile plot.

15. Select **Analyze > Fit Y by X**.

16. Select *Residual Shrinkage* for **Y, Response**.

17. Select *Hold Time* for **X, Factor**.

18. Click **OK**.

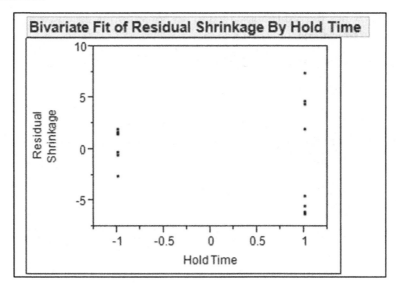

The plot of the residuals against *Hold Time* indicates that there is less process variability when *Hold Time* is held at its low level. Although it does not impact the mean shrinkage, *Hold Time* should be kept at its low level to keep the part-to-part variability low. We may build a model that estimates a separate error variance term for the different levels of *Hold Time* using the Loglinear Variance personality of the fit model platform.

19. Select **Analyze > Fit Model**.

20. Remove all terms from the Construct Model Effects area.

21. Select *Temperature, Screw,* and *Hold Time* under **Select Columns**.

22. Select **Macros > Full Factorial**.

23. Select **Personality > Loglinear Variance**.

24. Click the **Variance Effects** tab that now appears in the Construct Model Effects area.

25. Select *Hold Time* and click **Add**.

26. Check **Keep dialog open**.

27. Click **Run**.

Variance Model for Shrinkage

Likelihood Ratio Test for Equal Variance

Source	-2*LogLikelihood
Equal Variances Initially	71.415375535
After Fitted Variances	62.433648251
Difference: Chi-sq	8.9817272842
Degrees of Freedom	1
Prob > ChiSquare	0.0027269275

Variance Parameter Estimates

Term	Estimate	Std Error	ChiSquare	Prob>ChiSq	Lower 95% Limit	Upper 95% Limit
Hold Time	1.7882751	0.5	8.9817	0.0027*	0.72831	2.8482403
Residual	8.9686956	4.484348	4.0000	0.0455*	3.3661116	23.896267

Variance Effect Likelihood Ratio Tests

Source	Test Type	DF	ChiSquare	Prob>ChiSq
Hold Time	Likelihood	1	8.9817	0.0027*

The p-value of 0.0027 for the likelihood ratio test for equal variance indicates that the error variance is significantly different across holding times, and that the model is improved by including this effect.

28. Cube plots are not available when using the Loglinear Variance personality. Return to the fit model dialog and select **Standard Least Squares** for **Personality**.

29. Remove *Hold Time* from the **Random Effects** tab of the Construct Model Effects area.

30. Click the red triangle next to Response Shrinkage and select **Factor Profiling > Cube Plots**.

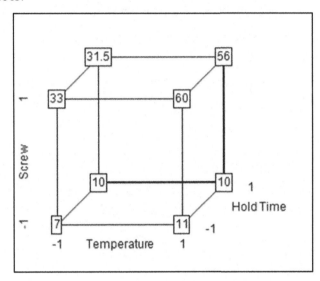

A relatively small mean part shrinkage will result from any combination of *Temperature* and *Hold Time* as long as *Screw* is set to its low level.

31. Select **Window > Close All**.

Example 8.5 A 2^{7-3} Design

1. Select **DOE > Screening Design**.

2. Enter 7 next to the **Continuous** field.

3. Click **Add**.

4. Click **Continue**.

5. Select the 16 run fractional factorial design (without blocking).

Design List			
Choose a design by clicking on its row in the list.			
Number Of Runs	Block Size	Design Type	Resolution - what is estimable
8		Fractional Factorial	3 - Main Effects Only
12		Plackett-Burman	3 - Main Effects Only
16		Fractional Factorial	4 - Some 2-factor interactions
16	8	Fractional Factorial	4 - Some 2-factor interactions
16	4	Fractional Factorial	4 - Some 2-factor interactions
16	2	Fractional Factorial	4 - Some 2-factor interactions
32		Fractional Factorial	4 - Some 2-factor interactions
32	16	Fractional Factorial	4 - Some 2-factor interactions
32	8	Fractional Factorial	4 - Some 2-factor interactions
32	4	Fractional Factorial	4 - Some 2-factor interactions
32	2	Fractional Factorial	4 - Some 2-factor interactions
64		Fractional Factorial	5+ - All 2-factor interactions
64	32	Fractional Factorial	5+ - All 2-factor interactions
64	16	Fractional Factorial	5+ - All 2-factor interactions

6. Click **Continue**.

If desired, the generating rules may be changed.

Display and Modify Design			
Change Generating Rules			
Factors	X5	X6	X7
X1	☐	☑	☑
X2	☑	☐	☑
X3	☑	☑	☐
X4	☑	☑	☑

Output Options	
Run Order:	Randomize
Make JMP Table from design plus	
Number of Center Points:	0
Number of Replicates:	0

7. Click **Make Table** to complete the process.

8. Select **Window > Close All**.

Example 8.6 A 2^{8-3} Design in Four Blocks

1. Select **DOE > Screening Design**.

2. In the **Continuous** field, enter 8.

3. Click **Add**.

4. Click **Continue**.

5. Select the 32 run fractional factorial design in 4 blocks of size 8.

Number Of Runs	Block Size	Design Type	Resolution - what is estimable
32	8	Fractional Factorial	4 - Some 2-factor interactions

6. Click **Continue**.

7. Click **Make Table**.

8. Open Jet-Engine.jmp. The responses are saved in this data table.

9. *The Screening platform is not appropriate for designs with categorical factors containing more than two levels, including blocking factors.* We will first remove the blocking effect by fitting the response against the block effect. The residuals from this analysis can then be used with the screening platform and the continuous factors.

   Warning: As a result, the Screening platform will not recognize that certain factors are confounded with the block effect. These alias chains should be tracked manually.

10. From the red triangle next to **Model**, click **Run Script**.

11. Select *ln(std_dev)* and click **Y**.

12. Select *Blocks* click **Add**.

13. Click **Run**.

14. Click the red triangle next to Response ln(std_dev) and select **Save Columns > Residuals**.

15. Return to the data table and choose **Analyze > Modeling > Screening**.

16. Select *Residual ln(std_dev)* and click **Y**.

17. Select *A* through *H* (do not include *Blocks*) and click **X**.

18. Click **OK**.

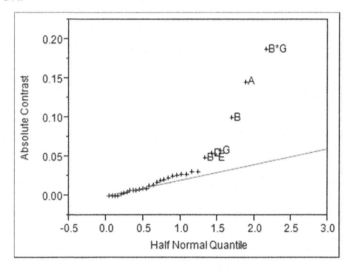

Contrasts

Term	Contrast		Lenth t-Ratio	Individual p-Value	Simultaneous p-Value	Aliases
A	0.145131		7.19	<.0001*	0.0010*	B*G*D, B*F*C
B	-0.100270		-4.97	0.0007*	0.0109*	A*G*D, A*F*C
G	0.058041		2.87	0.0110*	0.2304	A*B*D, D*F*C
D	0.054066		2.68	0.0159*	0.3013	A*B*G, G*F*C
F	-0.019355		-0.96	0.3251	1.0000	A*B*C, G*D*C
C	-0.012882		-0.64	0.5369	1.0000	A*B*F, G*D*F
H	0.007084		0.35	0.7336	1.0000	
E	-0.000253		-0.01	0.9896	1.0000	
A*B	-0.002940		-0.15	0.8847	1.0000	G*D, F*C
A*G	0.026440		1.31	0.1854	0.9907	B*D, C*H*E
B*G	-0.187058		-9.26	<.0001*	<.0001*	A*D, F*H*E
A*F	-0.022510		-1.11	0.2537	0.9994	B*C, D*H*E
B*F	-0.031032		-1.54	0.1263	0.9482	A*C, G*H*E
G*F	0.017261		0.85	0.3766	1.0000	D*C, B*H*E
D*F	-0.007334		-0.36	0.7251	1.0000	G*C, A*H*E
A*H	-0.025211		-1.25	0.2064	0.9945	D*F*E, G*C*E
B*H	0.006544		0.32	0.7522	1.0000	G*F*E, D*C*E
G*H	0.002446		0.12	0.9034	1.0000	B*F*E, A*C*E
D*H	0.007844		0.39	0.7057	1.0000	A*F*E, B*C*E
F*H	-0.014040		-0.70	0.4730	1.0000	B*G*E, A*D*E
C*H	0.030398		1.51	0.1329	0.9587	A*G*E, B*D*E
A*E	0.004019		0.20	0.8420	1.0000	D*F*H, G*C*H
B*E	0.049253		2.44	0.0244*	0.3867	G*F*H, D*C*H
G*E	-0.026855		-1.33	0.1807	0.9889	B*F*H, A*C*H
D*E	0.008541		0.42	0.6808	1.0000	A*F*H, B*C*H
F*E	-0.009041		-0.45	0.6630	1.0000	B*G*H, A*D*H
C*E	0.019910		0.99	0.3102	0.9999	A*G*H, B*D*H
H*E	0.000000		0.00	1.0000	1.0000	B*G*F, A*D*F, A*G*C, B*D*C
A*G*F	-0.027167		-1.35	0.1762	0.9866	B*D*F, B*G*C, A*D*C
A*B*H	0.000000		0.00	1.0000	1.0000	G*D*H, F*C*H
A*B*E	0.000000		0.00	1.0000	1.0000	G*D*E, F*C*E

Notice that the *H*E* interaction is confounded with blocks, but that the screening report does not report this since blocks were not included as an input. The largest effects are *A*, *B*, and *AD+BG+FHE*. Suppose subject matter knowledge suggests that the appropriate interaction is likely *AD*. We will now fit the reduced model with *A*, *B*, *AD*, and *D* (to preserve hierarchy).

19. From the red triangle next to **Model**, click **Run Script**.

20. Select *Blocks*, *A*, *B*, and *D* by holding down the *Ctrl* key while clicking each term in the Select Columns area.

21. Click **Add**.

22. Select *A* and *D* under **Select Columns**.

23. Click **Cross**.

24. Select *ln(std_dev)* for **Y**.

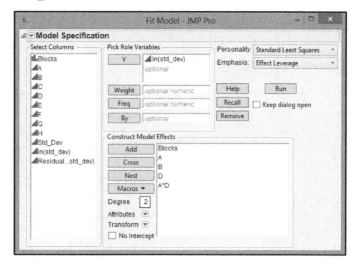

25. Click **Run**.

Effect Tests

Source	Nparm	DF	Sum of Squares	F Ratio	Prob > F
Blocks	3	3	0.0201391	0.3931	0.7591
A	1	1	0.6740199	39.4686	<.0001*
B	1	1	0.3217292	18.8395	0.0002*
D	1	1	0.0935397	5.4774	0.0279*
A*D	1	1	1.1197046	65.5665	<.0001*

All of the included factors are significant. The F test for the significance of the blocking factor should be analyzed with caution due to the randomization restriction within each block.

26. Click the red triangle next to Response ln(std_dev) and select **Factor Profiling > Interaction Plots**.

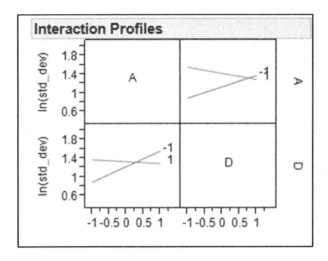

Factors *A* and *D* are strongly associated with the response. Running both factors at the low level minimizes *ln(std_dev)*. This may also be seen using the prediction profiler, which is demonstrated in Chapter 11.

27. Select **Window > Close All**.

Example 8.7 A Fold-Over 2^{7-4} Resolution III Design

1. Open Eye-Focus.jmp.

2. Click the red triangle next to the **Screening** script in the left panel and select **Run Script**.

Contrasts

Term	Contrast		Lenth t-Ratio	Individual p-Value	Simultaneous p-Value
B	19.1875		56.85	<.0001*	0.0002*
D	14.4375		42.78	<.0001*	0.0004*
A	10.3125		30.56	<.0001*	0.0007*
G	-1.2125		-3.59	0.0219*	0.1076
F	-0.3125		-0.93	0.3011	0.9467
E	-0.1375		-0.41	0.7282	1.0000
C	-0.1375		-0.41	0.7282	1.0000

A, B, and *D* are the largest effects. However, the aliasing of main effects with two-way interactions prevents us from simply concluding that *A, B,* and *D* are all significant. We will now use the full fold-over technique to separate the main effects from two-way interactions.

3.   Select **DOE > Augment Design**.

4.   Select *Time* for **Y, Response**.

5.   Select *A* through *G* for **X, Factor.**

6.   Click **OK**.

7.   Check the box for the option "Group new runs into separate block."

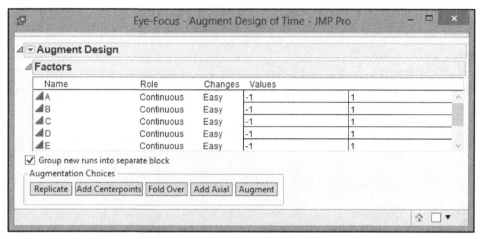

8.   Click **Fold Over**.

9.   Select all of the factors.

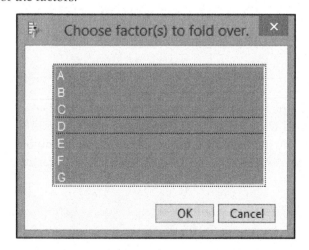

10. Click **OK**.

11. Click **Make Table**.

12. Open Eye-Focus-Foldover.jmp.

13. Select **Analyze > Modeling > Screening**. We may include the blocking factor in the Screening platform since it only appears at two levels. See Example 8.6 for further discussion.

14. Select *Time* for **Y**.

15. Select *Blocks* and *A* through *G* for **X**.

16. Click **OK**.

Contrasts

Term	Contrast			Lenth t-Ratio	Individual p-Value	Simultaneous p-Value	Aliases
B	19.0250			31.22	<.0001*	<.0001*	
D	14.6875			24.10	<.0001*	<.0001*	
Block	-1.0250			-1.68	0.1015	0.6854	
C	-0.9000			-1.48	0.1387	0.8191	
A	0.7375			1.21	0.2113	0.9509	
F	0.2500			0.41	0.7073	1.0000	
G	0.0625			0.10	0.9230	1.0000	
E	0.0625			0.10	0.9230	1.0000	
B*D	9.5750			15.71	<.0001*	0.0003*	Block*A, F*G, C*E
B*Block	-0.1625			-0.27	0.8051	1.0000	D*A, C*F, G*E
D*Block	0.2500			0.41	0.7073	1.0000	B*A, C*G, F*E
B*C	-0.5625			-0.92	0.3313	0.9982	Block*F, A*G, D*E
D*C	-1.2750			-2.09	0.0523	0.4270	A*F, Block*G, B*E
Block*C	-0.7625			-1.25	0.1982	0.9351	B*F, D*G, A*E
C*A	-0.2000			-0.33	0.7625	1.0000	D*F, B*G, Block*E

B and *D* are the largest effects. The third largest effect is the *B*D* interaction. Note that *B*D* is aliased with *F*G* and *C*E*. We ignore the reported *Blocks*A* alias, since we assume that there is no interaction between the factor effects and blocks. The principle of hierarchy makes it reasonable to assume that the *B*D* factor is the important effect in this alias chain.

17. Select **Window > Close All**.

Example 8.8 The Plackett-Burman Design

Plackett-Burman designs may be created in JMP using the Screening Design platform. Notice in the previous examples using the Screening Design platform, there was an option for selecting a Plackett-Burman design. The textbook uses forward stepwise selection in this example. While this method is available in JMP (via the Stepwise **Personality** for the Fit Model platform), the Screening platform provides a more straightforward solution.

1. Open Plackett-Burman.jmp.

2. Click the red triangle next to the **Screening** script in the left panel and select **Run Script**.

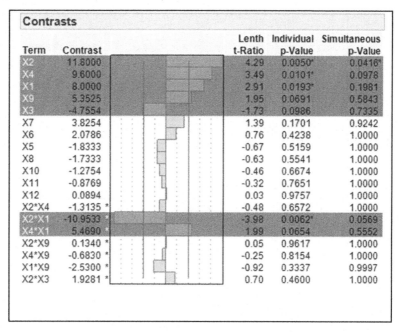

The Screening platform identifies the factors that were used to simulate the data (*X1, X2, X4, X2*X1,* and *X4*X1*), along with two additional factors (*X3* and *X9*). These last two factors represent Type I errors. This means thatthey have been incorrectly flagged as significant. In reference to screening experiments, Type I errors are not as severe as Type II errors (failing to flag an important effect as significant) since insignificant factors will easily be detected in future experimentation.

3. Select **Window > Close All**.

Section 8.7.2 Sequential Experimentation with Resolution IV Designs

1. Open Spin-Coater.jmp.

2. Click the red triangle next to the **Screening** script in the left panel and select **Run Script.**

Contrasts

Term	Contrast		Lenth t-Ratio	Individual p-Value	Simultaneous p-Value	Aliases
A	79.0625		11.55	<.0001*	0.0011*	B*C*E, E*F*D
B	-71.4375		-10.44	<.0001*	0.0019*	A*C*E, C*F*D
C	-40.4375		-5.91	0.0015*	0.0140*	A*B*E, B*F*D
E	27.8125		4.06	0.0072*	0.0595	A*B*C, A*F*D
F	-8.4375		-1.23	0.2060	0.9473	B*C*D, A*E*D
D	1.3125		0.19	0.8594	1.0000	B*C*F, A*E*F
A*B	61.0625		8.92	<.0001*	0.0024*	C*E
A*C	-9.6875		-1.42	0.1548	0.8594	B*E
B*C	9.5625		1.40	0.1583	0.8702	A*E, F*D
A*F	-10.1875		-1.49	0.1367	0.8155	E*D
B*F	1.8125		0.26	0.8088	1.0000	C*D
C*F	3.5625		0.52	0.6354	1.0000	B*D
E*F	4.3125		0.63	0.5675	1.0000	A*D
A*B*F	4.5625		0.67	0.5315	1.0000	C*E*F, A*C*D, B*E*D
A*C*F	0.5625		0.08	0.9395	1.0000	B*E*F, A*B*D, C*E*D

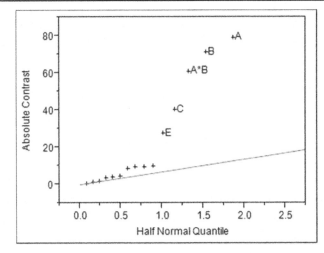

The Screening report identifies *A, B, C,* and *E* as the largest effects. The *A*B+C*E* alias chain is also important. However, without subject-matter knowledge, it is not possible to determine which of *A*B* or *C*E* (or both) are important. We will now build a fold-over design on factor *A* in order to break this alias chain in this Resolution IV design.

3.   Select **DOE > Augment Design**.

4.   Select *Thickness* for **Y, Response**.

5.   Select *A* through *F* and click **X, Factor**.

6.   Click **OK**.

7.   Check the option for "Group new runs into separate block."

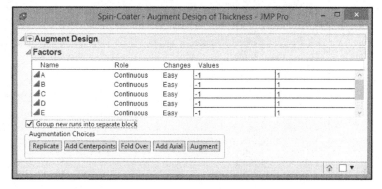

8.   Click **Fold Over**.

9.   Select *A*.

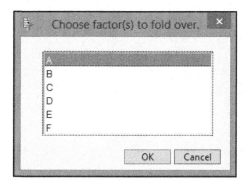

10.   Click **OK**.

11.   Click **Make Table**.

12. Open Spin-Coater-Foldover.jmp. The responses for the fold-over design have been preloaded into this data table.

13. Click the red triangle next to the **Screening** script in the left panel and select **Run Script**. We may include the blocking factor in the Screening platform since it only appears at two levels. See Example 8.6 for further discussion. The Screening platform should not be used for blocking factors with more than two levels.

Contrasts

Term	Contrast		Lenth t-Ratio	Individual p-Value	Simultaneous p-Value	Aliases
Block	6.7187		0.94	0.3405	1.0000	
A	76.4063		10.72	<.0001*	<.0001*	
B	-72.2812		-10.14	<.0001*	<.0001*	B*E*F
C	-35.8438		-5.03	0.0002*	0.0100*	A*E*F
E	35.4687		4.98	0.0002*	0.0110*	
F	-5.4687		-0.77	0.4376	1.0000	A*B*F
D	2.8437		0.40	0.6997	1.0000	A*B*E
A*B	13.0313		1.83	0.0766	0.8050	Block*B*E*F
A*C	-9.5313		-1.34	0.1842	0.9900	Block*A*E*F
B*C	11.0312		1.55	0.1267	0.9440	E*F
A*E	-1.4688		-0.21	0.8398	1.0000	
B*E	-0.1562		-0.02	0.9853	1.0000	B*C*E*F
C*E	48.0312		6.74	<.0001*	0.0012*	A*C*E*F
A*F	-1.6562		-0.23	0.8188	1.0000	Block*A*B*F
B*F	-6.5937		-0.93	0.3509	1.0000	B*F
C*F	-4.7812		-0.67	0.4990	1.0000	A*F
E*F	-0.3437		-0.05	0.9647	1.0000	A*B*C*F
A*D	4.6562		0.65	0.5381	1.0000	Block*A*B*E
E*D	-8.5313		-1.20	0.2313	0.9983	A*B*C*E
A*B*C	-7.6563		-1.07	0.2815	0.9996	Block*E*F
A*B*E	-4.5938		-0.64	0.5436	1.0000	
A*C*E	0.8437		0.12	0.9101	1.0000	
B*C*E	2.6562		0.37	0.7174	1.0000	C*E*F
A*B*F	8.2188		1.15	0.2492	0.9990	Block*B*F
A*C*F	5.2812		0.74	0.4538	1.0000	Block*A*F
A*E*F	-1.5312		-0.21	0.8319	1.0000	
B*E*F	-4.7188		-0.66	0.5317	1.0000	B*C*F
C*E*F	-3.6563		-0.51	0.6235	1.0000	A*C*F
A*E*D	-2.9688		-0.42	0.6881	1.0000	
A*B*E*F	8.3438		1.17	0.2422	0.9988	Block*B*C*F
A*C*E*F	8.4063		1.18	0.2392	0.9986	Block*A*C*F

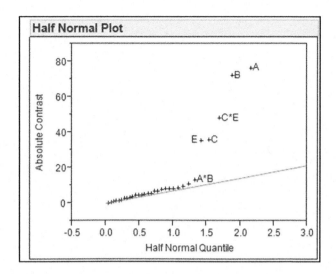

This plot reveals that $C*E$ is the more significant effect in the $A*B+C*E$ alias chain, but that the $A*B$ interaction may still be significant.

14. Scroll down to the bottom of the Screening platform and click **Run Model**. This fits a model using the effects highlighted on the **Screening** platform.

Parameter Estimates

Term	Estimate	Std Error	t Ratio	Prob>|t|
Intercept	4449.8438	5.767206	771.58	<.0001*
A	76.40625	5.767206	13.25	<.0001*
B	-72.28125	5.767206	-12.53	<.0001*
C	-35.84375	5.767206	-6.22	<.0001*
E	35.46875	5.767206	6.15	<.0001*
A*B	13.03125	5.767206	2.26	0.0328*
C*E	48.03125	5.767206	8.33	<.0001*

Both the $A*B$ and $C*E$ interactions are significant, though the $C*E$ interaction appears to have a much larger effect on the response.

15. Select **Window > Close All**.

Three-Level and Mixed-Level Factorial and Fractional Factorial Designs

Three-level factorial designs are introduced in this chapter. These designs are often used in experiments where at least one of the factors is qualitative (categorical) and naturally takes on three levels. Including three levels of each factor allows for the calculation of quadratic effects in quantitative factors. However, response surface designs (Chapter 11) are often more efficient for modeling quadratic relationships. The size of 3^k designs increases rapidly with k. For example, a 2^4 design requires 16 runs, whereas a 3^4 design requires 81 runs. If desired, the 2^k designs may be efficiently augmented with center runs to test for lack of fit due to the presence of an omitted quadratic effect.

Given the large number of runs needed for three-level factorial designs, it is often necessary to run these experiments in blocks. The 3^k designs may be confounded in 3^p incomplete blocks, where $p<k$. Likewise, fractional replications of these designs are often of interest. However, some of these fractional designs have complex alias structures.

Analysis of these designs is possible through the JMP Fit Model platform. This platform is also capable of analyzing mixed-level and non-regular fractional factorial designs, including "no-confounding" designs. To evaluate potential designs, JMP is capable of producing the correlation maps displayed in the textbook for visualizing confounding between effects. The last example of this chapter presents I- and D-optimal designs to satisfy a nonstandard blocking requirement, and uses tools provided by JMP to evaluate and compare the designs.

Example 9.1 The 3^3 Design

We first demonstrate how JMP may be used to create a 3^3 factorial design.

1. Select **DOE > Full Factorial Design**.

2. Double-click the Response Name **Y** and change it to *Syrup Loss*.

3. Click **Maximize** under Goal and select **Minimize**.

4. In the Factors report, click the drop-down menu **Categorical** and select **3 Level**. Repeat this step twice for a total of 3 three-level categorical factors.

5. Change the Name for the factors to *Nozzle Type, Speed,* and *Pressure*.

6. Change the Values for the factors to match the table below.

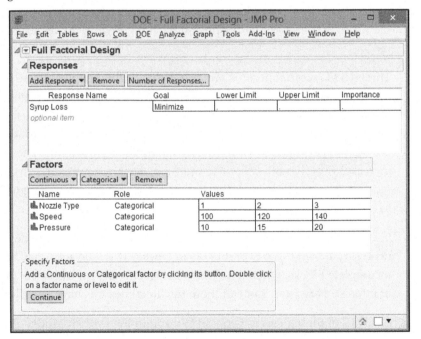

7. Click **Continue**.

8. Change **Number of Replicates** to 1. This replicates the original 27 runs once for a total of 54 runs.

9. Click **Make Table**.

10. Open Syrup-Loss.jmp. The results for *Syrup Loss* have been loaded into this data table.

11. From the red triangle next to **Model**, click **Run Script**. By default, JMP only includes two-way interaction effects in the Construct Model Effects area.

12. Select *Nozzle Type, Speed,* and *Pressure* in the Select Columns area and then click **Cross**.

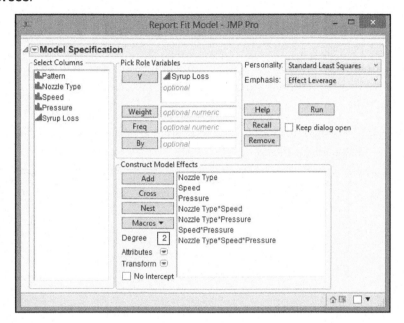

13. Click **Run**.

Analysis of Variance

Source	DF	Sum of Squares	Mean Square	F Ratio
Model	26	162587.33	6253.36	14.6620
Error	27	11515.50	426.50	Prob > F
C. Total	53	174102.83		<.0001*

Effect Tests

Source	Nparm	DF	Sum of Squares	F Ratio	Prob > F
Nozzle Type	2	2	993.778	1.1650	0.3271
Speed	2	2	61190.333	71.7354	<.0001*
Nozzle Type*Speed	4	4	6300.889	3.6934	0.0159*
Pressure	2	2	69105.333	81.0145	<.0001*
Nozzle Type*Pressure	4	4	7513.889	4.4044	0.0072*
Speed*Pressure	4	4	12854.333	7.5348	0.0003*
Nozzle Type*Speed*Pressure	8	8	4628.778	1.3566	0.2595

The three-factor interaction is not significant. However, all three of the two-factor interactions are important. In practice, three-factor and higher interactions are rarely significant.

14. Click the red triangle next to Response Syrup Loss and select **Factor Profiling > Interaction Plots**.

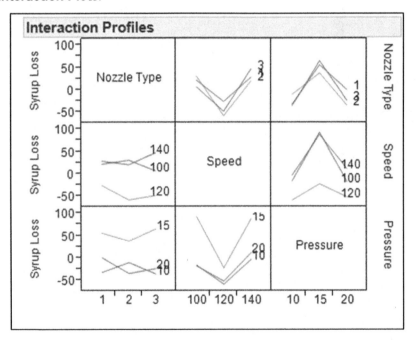

The 120 level of *Speed* appears to minimize syrup loss, along with the 10 or 20 psi settings for *Pressure*. This may be seen more easily using the Profiler.

15. Click the red triangle next to Response Syrup Loss and select **Factor Profiling > Profiler**.

We have already specified to JMP that *Syrup Loss* should be minimized in Step 3. This preference can be verified by clicking the red triangle next to Prediction Profiler and selecting **Set Desirabilities**.

16. Click the red triangle next to Prediction Profiler and select **Maximize Desirability**.

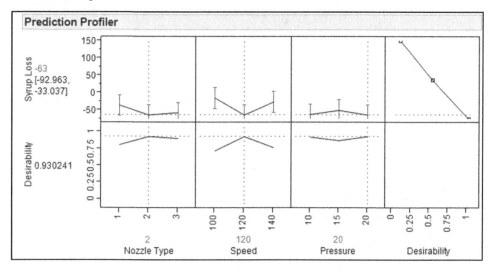

The minimum predicted *Syrup Loss* of -63 occurs with *Nozzle Type* 2 at *Speed* 120 with *Pressure* 20.

17. Select **Window > Close All**.

Example 9.2 The 3^2 Design Confounded in 3 Blocks

1. Open Confounded-Blocks.jmp.

Note: This design cannot be created using the Full Factorial in JMP. However, this design can easily be created using the Custom Design platform.

2. From the red triangle next to **Model**, click **Run Script**.

3. Click **Run**.

Analysis of Variance

Source	DF	Sum of Squares	Mean Square	F Ratio
Model	6	142.66667	23.7778	16.4615
Error	2	2.88889	1.4444	Prob > F
C. Total	8	145.55556		0.0584

Effect Tests

Source	Nparm	DF	Sum of Squares	F Ratio	Prob > F
Block	2	2	10.88889	3.7692	0.2097
A	2	2	131.55556	45.5385	0.0215*
B	2	2	0.22222	0.0769	0.9286

This analysis matches the results of Table 9.4 in the textbook. Since there is only one replicate, the *A*B* effect is treated as an estimate of error and is used in the denominator of the F tests.

4. Select **Window > Close All**.

Example 9.3 The Spin Coating Experiment

1. Open Photoresist-Material.jmp.

Note: This is a 6-factor, 16-run orthogonal design constructed using a main-effects only model in the Custom Design platform in JMP.

2. Click the red triangle next to the **Screening** script in the left panel and select **Run Script**.

Contrasts

Term	Contrast		Lenth t-Ratio	Individual p-Value	Simultaneous p-Value
A	112.875		7.21	0.0009*	0.0066*
B	-50.375		-3.22	0.0167*	0.1423
C	-34.250		-2.19	0.0504	0.3765
E	21.500		1.37	0.1653	0.8826
F	-11.250		-0.72	0.4468	1.0000
D	1.500		0.10	0.9257	1.0000
A*B	-4.875		-0.31	0.7729	1.0000
A*C	-1.237 *		-0.08	0.9383	1.0000
B*C	10.430 *		0.67	0.5315	1.0000
A*E	2.298 *		0.15	0.8876	1.0000
B*E	-17.501 *		-1.12	0.2491	0.9780
C*E	38.714 *		2.47	0.0353*	0.2597
C*F	-1.768 *		-0.11	0.9140	1.0000
E*F	14.875		0.95	0.3191	0.9978
C*D	10.125		0.65	0.5531	1.0000

The effects *A*, *B*, *C*, and *C*E* are relatively large. If we were to fit a reduced model with these effects (along with *E* to preserve hierarchy), we would find that *E* is significant.

3. The textbook fits this model using stepwise regression. To replicate the results in the book, from the red triangle next to **Model**, click **Run Script** . The Construct Model Effects area has been populated with the main effects.

4. Select *A* through *F* in the Select Columns area and click **Macros > Factorial to Degree**. This will add the two-way interactions to the Construct Model Effects area.

5. Change **Personality** to **Stepwise**.

6. Click **Run**.

7. Change **Stopping Rule** to **P-value Threshold**.

8. Enter 0.05 for **Prob to Enter**.

9. Enter 0.10 for **Prob to Leave**.

10. Change **Direction** to **Mixed**.

Note: The textbook uses Forward stepwise regression while we use Mixed stepwise regression; both provide the same results.

Stepwise Regression Control

Stopping Rule: | P-value Threshold |

| Prob to Enter | 0.05 |
| Prob to Leave | 0.1 |

Direction: | Mixed |

Rules: | Combine |

11. Click **Go**.

Current Estimates

Lock	Entered	Parameter	Estimate	nDF	SS	"F Ratio"	"Prob>F"
☑	☑	Intercept	4462.875	1	0	0.000	1
☐	☑	A	85.5	1	77976	54.068	2.44e-5
☐	☑	B	-77.75	1	64480.67	44.711	5.45e-5
☐	☑	C	-34.25	2	42749.5	14.821	0.00102
☐	☐	D	0	1	36	0.023	0.88402
☐	☑	E	21.5	2	31376.5	10.878	0.0031
☐	☐	F	0	1	2025	1.470	0.25619
☐	☐	A*B	0	1	380.25	0.244	0.63335
☐	☐	A*C	0	1	481.3333	0.311	0.59081
☐	☐	A*D	0	2	3641.333	1.351	0.31223
☐	☐	A*E	0	1	120.3333	0.076	0.78939
☐	☐	A*F	0	2	4908	2.064	0.18938
☐	☐	B*C	0	1	65.33333	0.041	0.84412
☐	☐	B*D	0	2	937.3333	0.278	0.76429
☐	☐	B*E	0	1	3675	3.078	0.11327
☐	☐	B*F	0	2	2041.333	0.660	0.54309
☐	☐	C*D	0	2	1676.25	0.526	0.61004
☐	☑	C*E	54.75	1	23980.5	16.628	0.00222
☐	☐	C*F	0	2	2075	0.672	0.5372
☐	☐	D*E	0	2	86	0.024	0.97636
☐	☐	D*F	0	0	0	.	.
☐	☐	E*F	0	2	5565.25	2.514	0.14223

The stepwise procedure has isolated the same effects as the Screening platform, in addition to the *E* effect.

12. We will now produce a correlation matrix for the effects in this design (see p. 423 of the textbook). Select **DOE > Evaluate Design**.

13. Select *A* through *F* for **X, Factor**.

14. Select *Thickness* for **Y, Response**.

15. Click **OK**.

16. Scroll down and click the gray triangle next to Color Map on Correlations to expand the report.

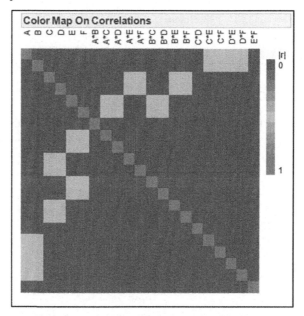

We can see that none of the main effects or two-way interactions are completely confounded with each other (areas in blue). The main effects are partially confounded with two-way interactions (areas in gray).

17. Select **Window > Close All**.

Example 9.4 An Experiment with Unusual Blocking Requirements

1. Select **DOE > Custom Design**.

2. Click the red triangle next to Custom Design and select **Optimality Criterion > Make I-Optimal Design**.

3. In the Factors report, enter 6 for **Add N Factors**.

4. Select **Add Factor > Continuous**.

5. Select **Add Factor > Blocking > 3 runs per block**.

Factors

Add N Factors [1]

Name	Role	Changes	Values	
X1	Continuous	Easy	-1	1
X2	Continuous	Easy	-1	1
X3	Continuous	Easy	-1	1
X4	Continuous	Easy	-1	1
X5	Continuous	Easy	-1	1
X6	Continuous	Easy	-1	1
X7	Blocking	Easy	1	

6. Click **Continue**.

7. Select the minimum number of runs, 12, in the Design Generation report.

Design Generation

Number of Center Points: [0]
Number of Replicate Runs: [0]

Number of Runs:
- ⊙ Minimum   12
- ○ Default   18
- ○ User Specifie[12]

8. Click **Make Design**.

Design

Run	X1	X2	X3	X4	X5	X6	X7
1	1	1	1	-1	-1	1	3
2	1	-1	-1	-1	1	1	4
3	1	1	-1	1	1	-1	3
4	1	-1	-1	1	-1	1	2
5	-1	1	-1	1	-1	1	4
6	-1	-1	-1	0	-1	-1	1
7	-1	1	1	-1	1	-1	2
8	-1	-1	1	1	1	1	3
9	1	-1	1	1	-1	-1	4
10	1	1	1	1	1	1	1
11	-1	0	0	-1	0	-1	2
12	0	-1	-1	-1	-1	0	1

Because an algorithm is used to generate this design, it does not represent a unique solution to the problem. There will likely be multiple designs that produce roughly the same value for the optimality criterion. Different designs may be obtained by clicking the red triangle next to Custom Design and selecting **Set Random Seed**. Different seeds will likely lead to different designs.

9. Click the gray triangle next to Design Evaluation to expand the report.

10. Click the gray triangle next to Fraction of Design Space Plot to expand the report.

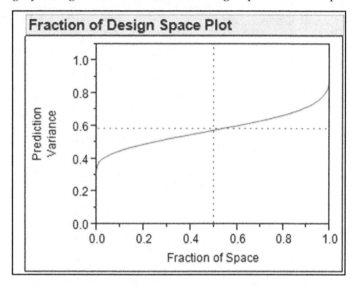

The Fraction of Design Space Plot provides a tool for comparing the prediction variance over the design space for different designs. Ideally, the prediction variance will be small and constant throughout the design space. See page 285 of the textbook for further discussion.

11. Click the gray triangle next to Color Map on Correlations to expand the report.

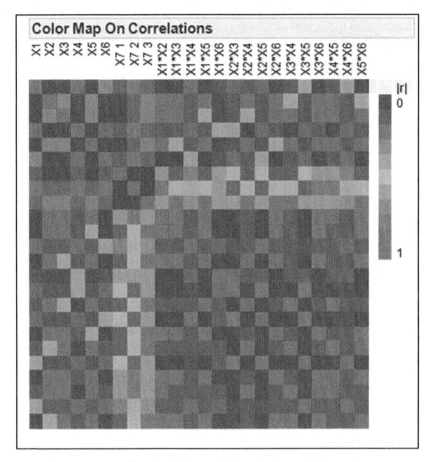

There is some partial confounding between main effects, and stronger partial confounding between main effects and two-way interactions. The strongest confounding occurs between pairs of two-way interactions.

12. Click the gray triangle next to Design Diagnostics to expand the report.

Design Diagnostics	
I Optimal Design	
D Efficiency	88.14938
G Efficiency	91.10217
A Efficiency	83.01508
Average Variance of Prediction	0.581921
Design Creation Time (seconds)	0.083333

This I-optimal design was built to minimize the Average Variance of Prediction. Next, we will build a D-optimal design for the same experiment that will maximize D Efficiency. The various design criteria are discussed in the textbook.

13. Click **Make Table**.

14. Return to the DOE – Custom Design window.

15. Click **Back**.

16. Click the red triangle next to Custom Design and select **Optimality Criterion > Make D-Optimal Design**.

17. Click **Make Design**.

Design								
Run	X1	X2	X3	X4	X5	X6	X7	Y
1	1	1	1	1	-1	-1	4	.
2	-1	1	1	-1	-1	-1	1	.
3	-1	1	-1	-1	1	1	3	.
4	-1	-1	-1	1	-1	-1	3	.
5	1	-1	1	-1	-1	1	3	.
6	1	1	-1	-1	-1	-1	2	.
7	1	-1	-1	1	1	-1	1	.
8	1	-1	-1	-1	-1	1	4	.
9	-1	-1	1	-1	1	-1	4	.
10	1	1	1	1	1	1	2	.
11	-1	1	-1	1	-1	1	1	.
12	-1	-1	1	1	-1	1	2	.

Unlike with the I-optimal design, the D-optimal criterion will always place the runs on the boundary of the design space. Hence, there are no 0s in the above matrix.

18. Click the gray triangle next to Design Evaluation to expand the report.

19. Click the gray triangle next to Fraction of Design Space Plot to expand the report.

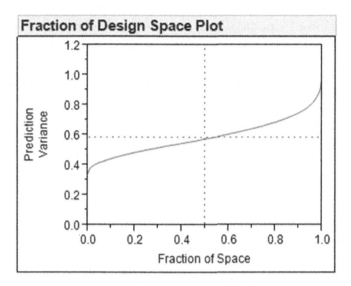

The differences between this FDS plot and the one created for the I-optimal design are relatively small. However, it is possible to see that the prediction variance for this design is generally larger, especially at the extremes of the plot.

20. Click the gray triangle next to Color Map on Correlations to expand the report.

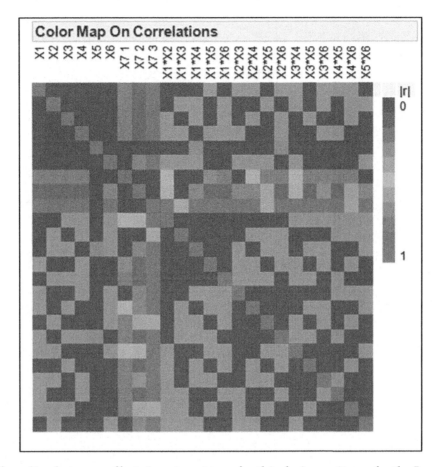

The confounding between effects is not as strong for this design as it was for the I-optimal design. Whereas the I-optimal criterion minimizes the average prediction variance, the D-optimal criterion minimizes the variance of the model parameter estimates. These facts are evident in the better performance of the I-optimal design on the FDS plot and in the better performance of the D-optimal design in the Color Map on Correlations plot.

21. Click the gray triangle next to Design Diagnostics to expand the report.

Design Diagnostics	
D Optimal Design	
D Efficiency	92.88393
G Efficiency	84.33797
A Efficiency	85.88957
Average Variance of Prediction	0.59127
Design Creation Time (seconds)	0.066667

The Average Variance of prediction has increased from 0.5819 to 0.5913 when switching from an I- to a D-optimal design, but the D Efficiency has increased from 88.15 to 92.88.

22. Click **Make Table**.

23. Select **Window > Close All**.

Fitting Regression Models

Linear regression provides the ability to analyze the majority of the models that are presented in this book. These models are often fit with the method of least squares, though the estimators may also be derived from a maximum likelihood framework. The two methods produce equivalent point estimates for the model parameters (with a small caveat for the estimates of variance components). However, the least squares method tends to produce more precise confidence intervals for the parameter estimates when the sample size is small, since the maximum likelihood method relies on asymptotic approximations. The maximum likelihood method is extremely useful for the estimation of models with random effects, as well as for models with non-normal response distributions. The maximum likelihood estimates may be obtained by setting the Personality option of the Fit Model platform to Generalized Linear Model. This option will be discussed further in Chapters 13 and 15.

With linear regression methods, you will have no difficulty analyzing data sets with missing observations (as long as you are willing to assume that the data are missing at random) or experiments where the factor levels could not be set to their exact design levels. Furthermore, several diagnostics are available to assess the fit of regression models. We have already seen how the assumptions that the errors are independent and identically normally distributed may be validated by examining the residuals. In addition, each observation is associated with a leverage value. The leverage values are the diagonals of the hat matrix $\mathbf{X(X'X)^{-1}X'}$, where $\mathbf{X}$ is the design matrix. Observations with a leverage greater than $2*p/n$ (where p and n are the number of columns and rows, respectively, of $\mathbf{X}$) are likely outliers in the design space, and could potentially (but not necessarily) have a large impact on the regression coefficients. A related diagnostic, Cook's Distance, measures how much the predicted regression equation changes with the deletion of each individual observation. Rows with a Cook's Distance of greater than 1 are considered to be highly influential. These observations should be investigated, and should not simply be discarded.

Example 10.1 Multiple Linear Regression Model

1. Open Polymer-Viscosity.jmp.

2. From the red triangle next to **Model**, click **Run Script**.

3. Click **Run**.

Summary of Fit

RSquare	0.92697
RSquare Adj	0.915735
Root Mean Square Error	16.3586
Mean of Response	2348.563
Observations (or Sum Wgts)	16

Parameter Estimates

Term	Estimate	Std Error	t Ratio	Prob>|t|
Intercept	1566.0778	61.59184	25.43	<.0001*
Temperature	7.6212901	0.61843	12.32	<.0001*
Feed Rate	8.5848459	2.438684	3.52	0.0038*

Both *Temperature* and *Feed Rate* are significant at the 0.05 level, and both are positively correlated with *Viscosity*.

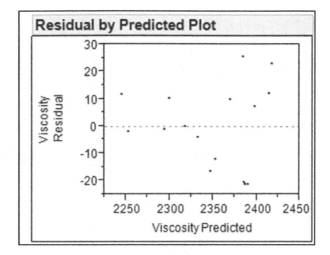

There is some evidence that the residual variance increases with the predicted *Viscosity*.

4. Hold down the *Alt* key and click the red triangle next to **Response Viscosity**.

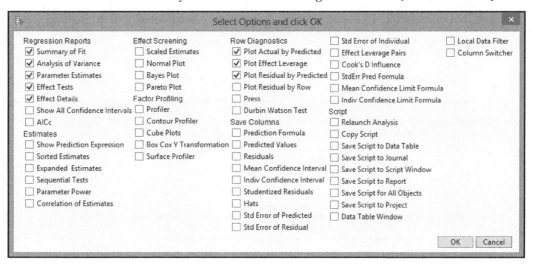

This provides a menu that enables you to save each of the following quantities back to the data table:

    a.   Predicted Values

    b.   Residuals

    c.   Studentized Residuals

      d. Hats

      e. Cook's D Influence

5. After checking each of these options, click **OK**.

6. Return to the Polymer-Viscosity data table and notice the new columns.

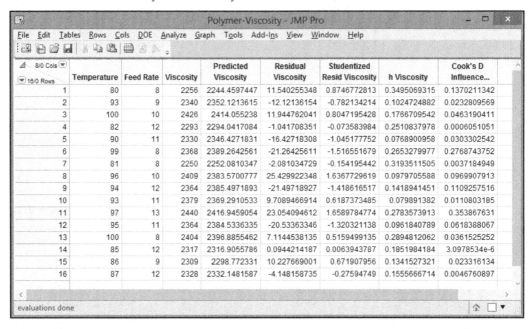

These diagnostics are discussed in Section 10.7 in the textbook.

7. Select **Analyze > Distribution**.

8. Select *Residual Viscosity* for **Y, Columns**.

9. Check **Histograms Only**.

10. Click **OK**.

11. Click the red triangle next to Residual Viscosity and select **Normal Quantile Plot**.

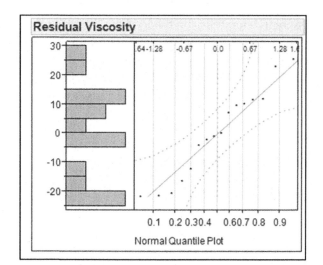

It appears to be reasonable to assume that the residual errors are normally distributed.

12. Return to the data table and select **Fit Y by X**.

13. Select *Residual Viscosity* for **Y, Response**.

14. Select *Temperature and Feed Rate* for **X, Factor**.

15. Click **OK**.

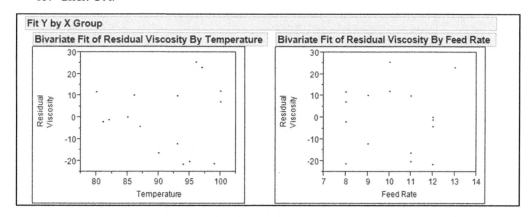

There is some indication that the error variance increases as *Temperature* increases. There is no apparent relationship between the residuals and *Feed Rate*.

Least squares regression relies on theory from linear algebra to produce parameter estimates. The matrix algebra occurs internally, although JMP also provides the ability to perform matrix operations explicitly through the JMP Scripting Language (JSL) interface. To demonstrate this capability, we will now work through a script that manually produces the regression estimates for the *Viscosity* example.

16. Open **Regression-Script.jsl** and read through the file. The green text following "//" is not processed by the JSL compiler: these are comments for the user. The script is included below for reference.

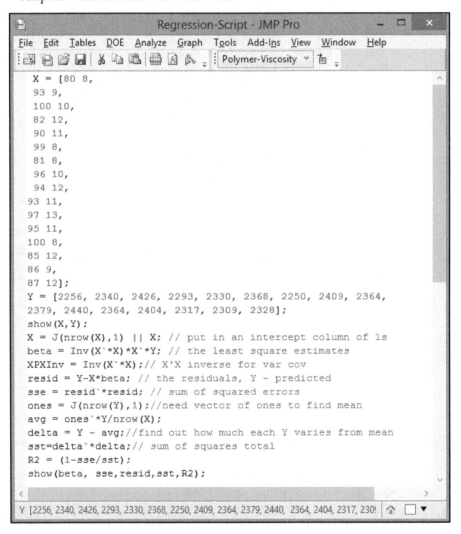

```
X = [80 8,
93 9,
100 10,
82 12,
90 11,
99 8,
81 8,
96 10,
94 12,
93 11,
97 13,
95 11,
100 8,
85 12,
86 9,
87 12];
Y = [2256, 2340, 2426, 2293, 2330, 2368, 2250, 2409, 2364,
2379, 2440, 2364, 2404, 2317, 2309, 2328];
show(X,Y);
X = J(nrow(X),1) || X; // put in an intercept column of 1s
beta = Inv(X`*X)*X`*Y; // the least square estimates
XPXInv = Inv(X`*X);// X'X inverse for var cov
resid = Y-X*beta; // the residuals, Y - predicted
sse = resid`*resid; // sum of squared errors
ones = J(nrow(Y),1);//need vector of ones to find mean
avg = ones`*Y/nrow(X);
delta = Y - avg;//find out how much each Y varies from mean
sst=delta`*delta;// sum of squares total
R2 = (1-sse/sst);
show(beta, sse,resid,sst,R2);
```

Y [2256, 2340, 2426, 2293, 2330, 2368, 2250, 2409, 2364, 2379, 2440, 2364, 2404, 2317, 230!

17. Select **Edit > Run Script** (or click the 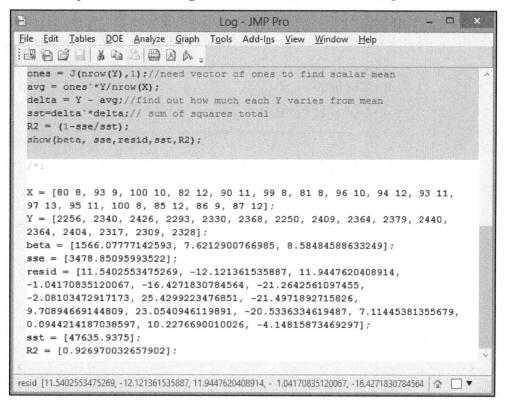 icon on the toolbar).

18. The output will be in the log file. To access it, select **View > Log**.

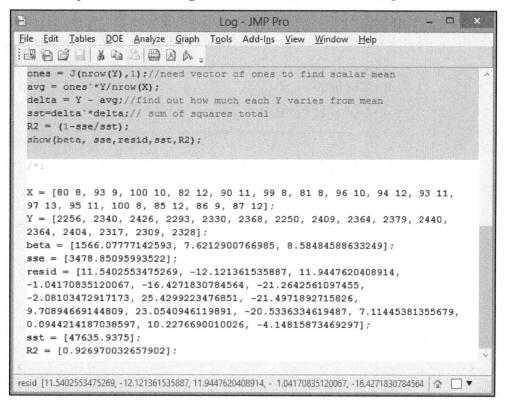

The parameter estimates reported in the array **beta** in the log file match those reported by the Fit Model platform.

19. Select **Window > Close All**.

Example 10.2 Regression Analysis of a 2^3 Factorial Design

We first demonstrate how JMP may be used to create a 2^3 factorial design.

1. Select **DOE > Full Factorial Design**.

2. Double-click the Response Name, *Y*, and change it to *Yield*.

3. In the Factors report, select **Continuous > 2 Level**. Repeat this step twice to create a total of three factors.

4. Double-click the factor *X1* and change its name to *Temp*. Change the low and high values to 120 and 160.

5. Double-click the factor *X2* and change its name to *Pressure*. Change the low and high values to 40 and 80.

6. Double-click the factor *X3* and change its name to *Conc*. Change the low and high values to 15 and 30.

7. Click **Continue**.

8. For **Number of Center Points**, enter 4.

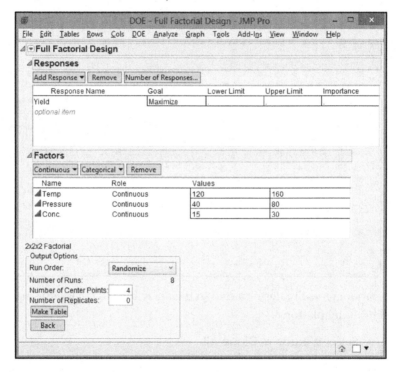

9. Click **Make Table**.

10. Open Process-Yield.jmp. The observed values of *Yield* have been loaded into this data table.

11. From the red triangle next to **Model**, click **Run Script**.

12. Select the two factor interaction terms, click **Remove**.

13. Click **Run**.

Parameter Estimates

Term	Estimate	Std Error	t Ratio	Prob>|t|
Intercept	51	0.966227	52.78	<.0001*
Temp(120,160)	5.625	1.183381	4.75	0.0014*
Pressure(40,80)	10.625	1.183381	8.98	<.0001*
Conc.(15,30)	1.125	1.183381	0.95	0.3696

Temp and *Pressure* are significant at the 0.05 level, but *Conc.* is not.

14. Click the red triangle next to Response Yield and select **Estimates > Show Prediction Expression**.

Prediction Expression

$$51$$

$$+ \; 5.625 * \left[\frac{(Temp - 140)}{20} \right]$$

$$+ \; 10.625 * \left[\frac{(Pressure - 60)}{20} \right]$$

$$+ \; 1.125 * \left[\frac{(Conc. - 22.5)}{7.5} \right]$$

This Prediction Expression illustrates how the factors are converted from natural to coded units.

15. Select **Window > Close All**.

Example 10.3 A 2^3 Factorial Design with a Missing Observation

1. Open Process-Yield-Missing.jmp.

2. From the red triangle next to **Model**, click **Run Script**.

3. Select the two factor interaction terms, click **Remove**.

4. Click **Run**.

Parameter Estimates

Term	Estimate	Std Error	t Ratio	Prob>|t|
Intercept	51.057692	1.107947	46.08	<.0001*
Temp(120,160)	5.7115385	1.401454	4.08	0.0047*
Pressure(40,80)	10.711538	1.401454	7.64	0.0001*
Conc.(15,30)	1.2115385	1.401454	0.86	0.4160

The parameter estimates are not substantially different from those obtained with all 12 runs in the previous example.

5. Select **Window > Close All**.

Example 10.4 Inaccurate Levels in Design Factors

1. Open Inaccurate-Levels.jmp.

2. From the red triangle next to **Model**, click **Run Script**.

3. Select the two factor interaction terms, click **Remove**.

4. Click **Run**.

Parameter Estimates

Term	Estimate	Std Error	t Ratio	Prob>|t|
Intercept	50.493938	1.034442	48.81	<.0001*
Temp(120,160)	5.4099296	1.253093	4.32	0.0026*
Pressure(40,80)	10.163124	1.225454	8.29	<.0001*
Conc.(15,30)	1.0725374	1.177118	0.91	0.3888

These estimates are similar to those obtained in Example 10.2. The inability of the experimenter to set the factors exactly at the specified levels does not seriously affect the results in this application. However, the standard errors of the parameter estimates are not identical with inaccurate levels because the design is no longer orthogonal.

5. Select **Window > Close All**.

Example 10.6 Tests on Individual Regression Coefficients

1. Open Polymer-Viscosity.jmp

2. Select **Analyze > Fit Model**.

3. Select *Viscosity* for **Y**.

4. Select *Temperature* and *Feed Rate*.

5. Click **Add**.

6. Click **Run**.

Effect Tests

Source	Nparm	DF	Sum of Squares	F Ratio	Prob > F
Temperature(80,100)	1	1	40641.418	151.8715	<.0001*
Feed Rate(8,13)	1	1	3316.244	12.3924	0.0038*

Both effects contribute significantly to the variability in *Viscosity*.

7. Select **Window > Close All**.

Example 10.7 Confidence Intervals on Individual Regression Coefficients

1. Open Polymer-Viscosity.jmp.

2. From the red triangle next to **Model**, click **Run Script**.

3. Click **Run**.

Analysis of Variance

Source	DF	Sum of Squares	Mean Square	F Ratio
Model	2	44157.087	22078.5	82.5046
Error	13	3478.851	267.6	Prob > F
C. Total	15	47635.938		<.0001*

4. Click the red triangle next to Response Viscosity and select **Regression Reports > Show All Confidence Intervals**.

Parameter Estimates

Term	Estimate	Std Error	t Ratio	Prob>|t|	Lower 95%	Upper 95%
Intercept	1566.0778	61.59184	25.43	<.0001*	1433.0167	1699.1388
Temperature	7.6212901	0.61843	12.32	<.0001*	6.2852541	8.9573261
Feed Rate	8.5848459	2.438684	3.52	0.0038*	3.3163898	13.853302

The 95% confidence interval for the *Temperature* coefficient is 6.29 to 8.96 while the interval for the *Feed Rate* coefficient is 3.32 to 13.85. Since both of these intervals do not include 0, we conclude that both effects significantly contribute to the variability in *Viscosity*. This check on the confidence intervals is equivalent to checking whether the *p*-values are less than or equal to 0.05.

5. Click the red triangle next to Response Viscosity and select **Row Diagnostics > Press**.

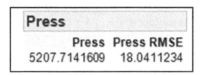

Press	
Press	**Press RMSE**
5207.7141609	18.0411234

Section 10.7.1 in the textbook discusses the prediction error sum of squares (PRESS) and shows how it may be used to evaluate the predictive performance of the model. The textbook reports a PRESS of 5207.7 for the *Viscosity* data. As expected, the PRESS of 5207.7 is greater than the SSE of 3478.9.

6. Select **Window > Close All**.

Response Surface Methods and Designs

The focus of many of the experimental designs introduced so far has been on factor screening. Screening designs are excellent for isolating the important factors from a set of several candidates, many of which may be unimportant. Once the screening experiments have identified the important factors to characterize the system, it may be necessary to find the settings of these factors that optimize one or multiple responses. Response surface methods provide us with a set of tools for these optimization problems.

Response surface methods ideally use explore the design space sequentially. At the start, a first order model may be centered on the current operating conditions. The path of steepest ascent (descent) is identified from this model, and individual experiments are run in steps along this path until the response no longer increases (or decreases). Another first-order model is fit, and the process is repeated. Once a first-order model indicates a lack of fit due to quadratic curvature, a second order model is fit by augmenting the existing design; central composite designs (CCDs) are often used at this point. The resulting second order response surface is then searched for its maximum (minimum) value. When multiple responses are of interest and when only a few factors are present, analyzing contour plots may be used to highlight acceptable regions of the design space. More generally, we maximize desirability functions of the responses subject to constraints.

A special class of response surface methods involves mixture experiments. In a mixture experiment, the factor levels being considered are proportions of that factor in a mixture of substances. The constraint that the sum of the proportions of each factor must equal 1 introduces challenges in the design and analysis of these experiments. Simplex lattice and simplex centroid designs are popular choices for these problems. Occasionally, other factor constraints are active (either in mixture or in ordinary response surface problems). Due to these constraints, the usual designs are no longer applicable and an optimal design must be created. Optimal designs may be created in JMP using the Custom Design and Mixture Design platforms.

Example 11.1 The Path of Steepest Ascent

We first demonstrate how JMP may be used to create a 2^2 factorial design with center points.

1. Select **DOE > Full Factorial Design**.

2. Double-click the Response Name, *Y*, and change it to *Yield*.

3. In the Factors area, select **Continuous > 2 Level**. Repeat this step, creating two factors, *X1* and *X2*.

4. Replace the low and high levels of *X1* with 30 and 40, respectively.

5. Replace the low and high levels of *X2* with 150 and 160, respectively.

6. Click **Continue**.

7. Enter 5 for **Number of Center Points**.

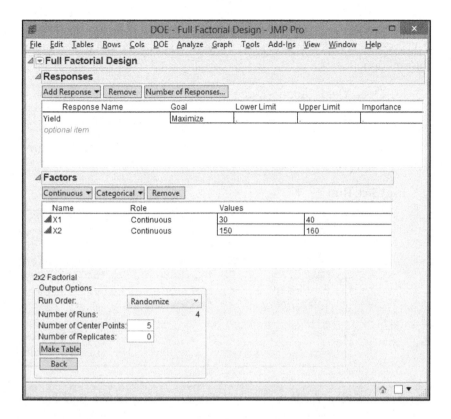

8. Click **Make Table**.

9. Open Yield-First-Model.jmp. The responses from the experiment have been preloaded into this data table.

10. From the red triangle next to **Model**, click **Run Script**.

11. Select the interaction *X1*X2* in the Construct Model Effects area and click **Remove**.

12. Click **Run**.

Parameter Estimates

Term	Estimate	Std Error	t Ratio	Prob>|t|
Intercept	40.444444	0.057288	705.99	<.0001*
X1(30,40)	0.775	0.085932	9.02	0.0001*
X2(150,160)	0.325	0.085932	3.78	0.0092*

The path of steepest ascent passes through the point ($X1$=0, $X2$=0) and has slope $0.325/0.775 = 0.42$. Table 11.3 from the textbook may be calculated manually using this information.

13. Open Yield-Second-Model.jmp.

14. From the red triangle next to **Model**, click **Run Script**.

15. Select the interaction $X1*X2$ in the Construct Model Effects area and click **Remove**.

16. Click **Run**.

Lack Of Fit

Source	DF	Sum of Squares	Mean Square	F Ratio
Lack Of Fit	2	10.908000	5.45400	102.9057
Pure Error	4	0.212000	0.05300	Prob > F
Total Error	6	11.120000		0.0004*

Max RSq
0.9868

The significance of the lack of fit test indicates that omitted model terms (either interaction or quadratic) are important. That is, the first order model does not provide an adequate approximation of the relationship between the predictors and the response variable.

Example 11.2 Central Composite Design

We will augment the second design from Example 11.1 to include axial runs.

1. Open Yield-Second-Model.jmp.

2. Select **DOE > Augment Design**.

3. Select *Yield* for **Y, Response**.

4. Select *X1* and *X2*, and choose **X, Factor**.

5. Click **OK**.

6. Click **Add Axial**.

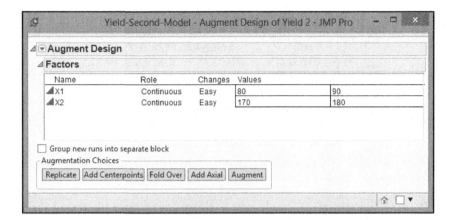

7. Enter an axial value of 1.414 and request 1 center point.

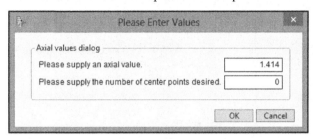

8. Click **OK**.

9. Click **Make Table**.

10. Open Yield-Axial-Points.jmp. The observed values of *Yield* have been loaded into this data table.

11. From the red triangle next to **Model**, click **Run Script**.

12. Remove the terms from the Construct Model Effects area.

13. Select *X1* and *X2* and select **Macros > Response Surface**. This will be the default behavior of designs created by the **DOE > Response Surface Design** platform. Since our response surface design resulted from the Augment Design platform, we need to use the **Response Surface** macro of the Fit Model platform.

14. Click **Run**.

Parameter Estimates

Term	Estimate	Std Error	t Ratio	Prob>|t|
Intercept	79.939955	0.119089	671.26	<.0001*
X1(80,90)	0.9950503	0.094155	10.57	<.0001*
X2(170,180)	0.5152028	0.094155	5.47	0.0009*
X1*X1	-1.376449	0.100984	-13.63	<.0001*
X1*X2	0.25	0.133145	1.88	0.1025
X2*X2	-1.001336	0.100984	-9.92	<.0001*

The pure quadratic terms are significantly related to *Yield*.

15. Click the gray triangle next to Response Surface to expand the section.

16. Click the gray triangle next to Canonical Curvature to expand the section.

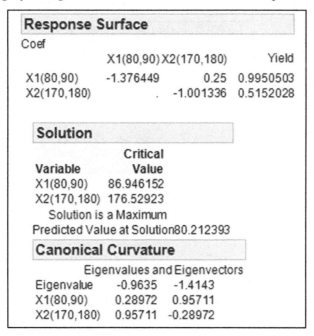

Response Surface

Coef

	X1(80,90)	X2(170,180)	Yield
X1(80,90)	-1.376449	0.25	0.9950503
X2(170,180)	.	-1.001336	0.5152028

Solution

Variable	Critical Value
X1(80,90)	86.946152
X2(170,180)	176.52923

Solution is a Maximum
Predicted Value at Solution 80.212393

Canonical Curvature

Eigenvalues and Eigenvectors

Eigenvalue	-0.9635	-1.4143
X1(80,90)	0.28972	0.95711
X2(170,180)	0.95711	-0.28972

The maximum *Yield* of 80.21 is predicted to result from setting *X1* = 86.95 and setting *X2* = 176.53. This solution is found via the canonical analysis provided via the Canonical Curvature report, as described in Section 11.3.2 of the textbook.

17. Click the red triangle next to Response Yield and select **Factor Profiling > Contour Profiler**.

18. Click the red triangle next to Contour Profiler and select **Contour Grid**.

19. Click **OK** to accept the default values for the contour plot.

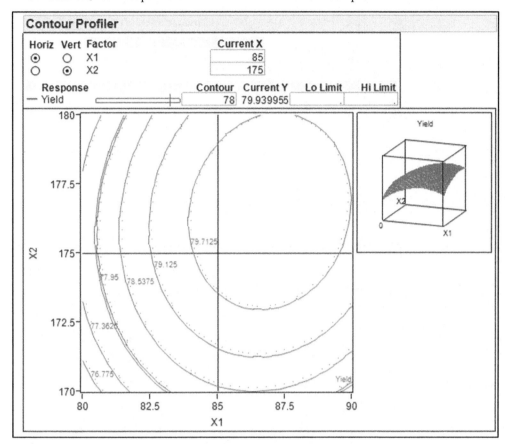

The maximum *Yield* occurs somewhere in the upper right quadrant of the above plot.

20. Click the red triangle next to Response Yield and select **Factor Profiling > Surface Profiler**.

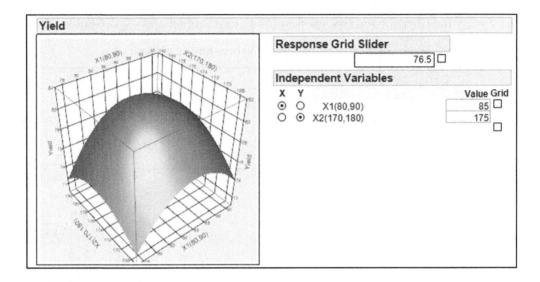

21. Scroll up to the section labeled Prediction Profiler.

22. Click the red triangle next to Prediction Profiler and select **Maximize Desirability**.

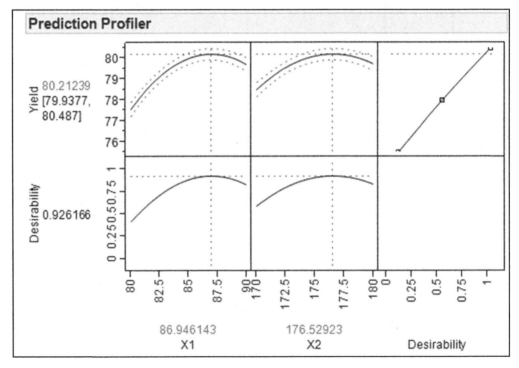

This is the same conclusion reached by the Response Surface report of the Fit Model platform. The prediction profiler is available any time that the Fit Model platform is used. It is also useful when multiple responses need to be optimized.

Note: The desirability limits can be saved as a column property for the response (for future use) by selecting **Save Desirabilities** in the menu under the red triangle next to Prediction Profiler.

    23. Select **Window > Close All**.

Section 11.3.4 Multiple Responses

    1. Open Multi-Response-CCD.jmp. This data table contains additional responses besides *Yield*, which was studied in the previous exercise. This data table was created using a rotatable CCD with the Response Surface Design platform.

    2. From the red triangle next to **Model**, click **Run Script**.

    3. Click **Run**.

Response Yield

Parameter Estimates

Term	Estimate	Std Error	t Ratio	Prob>|t|
Intercept	79.94	0.118959	672.00	<.0001*
X1(80,90)	0.9949747	0.094045	10.58	<.0001*
X2(170,180)	0.515165	0.094045	5.48	0.0009*
X1*X2	0.25	0.133	1.88	0.1022
X1*X1	-1.37625	0.100852	-13.65	<.0001*
X2*X2	-1.00125	0.100852	-9.93	<.0001*

Response Viscosity

Parameter Estimates

Term	Estimate	Std Error	t Ratio	Prob>|t|
Intercept	70	1.017546	68.79	<.0001*
X1(80,90)	-0.15533	0.80444	-0.19	0.8524
X2(170,180)	-0.948223	0.80444	-1.18	0.2770
X1*X2	-1.25	1.13765	-1.10	0.3082
X1*X1	-0.6875	0.862666	-0.80	0.4517
X2*X2	-6.6875	0.862666	-7.75	0.0001*

Response Weight				
Parameter Estimates				
Term	Estimate	Std Error	t Ratio	Prob>|t|
Intercept	3376	77.07072	43.80	<.0001*
X1(80,90)	205.10408	60.92976	3.37	0.0120*
X2(170,180)	177.35281	60.92976	2.91	0.0226*
X1*X2	-80	86.16769	-0.93	0.3841
X1*X1	-41.75	65.33989	-0.64	0.5432
X2*X2	58.25	65.33989	0.89	0.4023

We fit the same base model for each response (though with different parameter estimates for each response). The textbook fits a reduced model for the *Weight* response. If different base models are required for each response, the responses may be modeled individually. The prediction formula may then be saved and analyzed using the Contour Profiler or the Profiler platforms (under the **Graph** menu). See Example 12.2 for a demonstration.

4. Click the red triangle next to Least Squares Fit. Select **Profilers > Contour Profiler**.

5. Enter the required response limits as shown below.

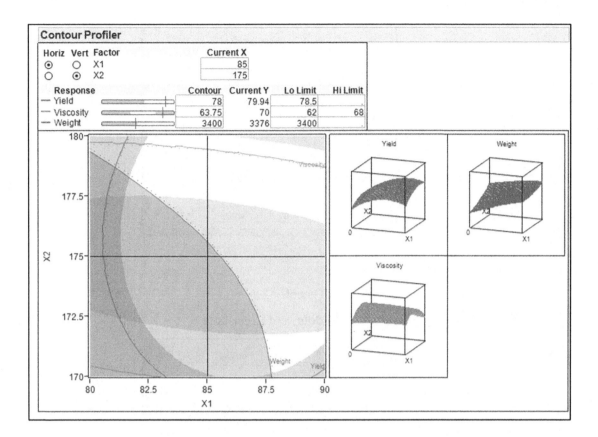

The white region (unshaded) in the contour profiler will lead to satisfactory operating conditions.

6. Scroll up to the section labeled Prediction Profiler.

7. Click the red triangle next to Prediction Profiler and select **Desirability Functions**.

8. Click the red triangle next to Prediction Profiler and select **Set Desirabilities**.

9. For *Yield*, select **Maximize** if it is not already selected.

10. Set **Importance** to 1. The **Importance** field may be decreased to indicate that a response is relatively less important than other responses. Setting the **Importance** of all responses to 1 treats them all equally.

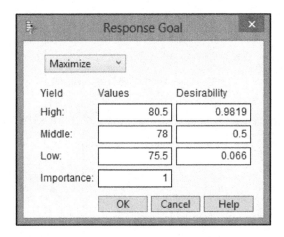

11. Click **OK**.

12. For *Viscosity*, select **Match Target**.

13. Enter 68 for **High**, 65 for **Middle**, and 62 for **Low**.

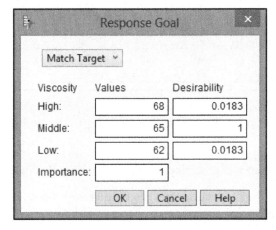

14. Click **OK**.

15. For *Weight*, select **Match Target**.

16. Enter 3400 for **High**, 3300 for **Middle**, and 3200 for **Low.**

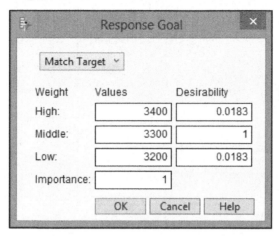

17. Click **OK**.

18. Click the red triangle next to Prediction Profiler and select **Maximize Desirability**.

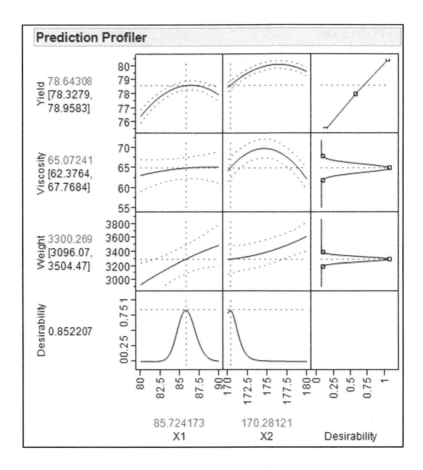

The recommended setting of *X1* = 85.72 and *X2* = 170.28 leads to a *Yield* of 78.64 with values of *Viscosity* and *Weight* that are within the specification limits.

Note: The desirability limits can be saved as a column property for the responses (for future use) by selecting **Save Desirabilities** in the menu under the red triangle next to Prediction Profiler.

19. Select **Window > Close All**.

Example 11.4 Space Filling Design with Gaussian Process Model

1. Select **DOE > Space Filling Design**.

2. Change the response name from *Y* to *Temperature*.

3. Change the name of factor *X1* to *x-axis* and enter values of 0.05 and 0.095.

4. Change the name of factor *X1* to *R-axis* and enter values 0 and 0.062.

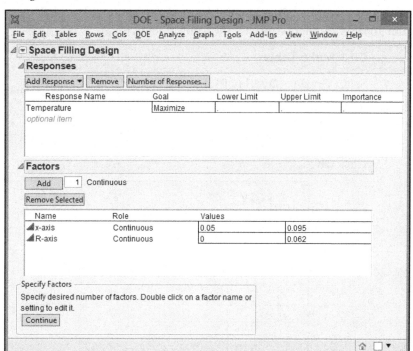

5. Click **Continue.**

6. Enter in **10** for **Number of Runs.**

7. Click **Sphere Packing**.

Factor Settings		
Run	x-axis	R-axis
1	0.06760	0.03591
2	0.06964	0.00000
3	0.05071	0.00000
4	0.05004	0.02606
5	0.06753	0.06200
6	0.09500	0.03873
7	0.08646	0.06200
8	0.09500	0.00000
9	0.05000	0.05215
10	0.08228	0.01941

This 10 run design resembles the one that appears in Table 11.16. Note that the algorithm will produce random points within some criteria and will randomize the run order, so your 10 runs will not be exactly the same as shown.

8. Click **Make Table**.

9. Select **Graph > Overlay Plot**.

10. Select *R-axis* for **Y.**

11. Select *x-axis* for **X**.

12. Click **OK**.

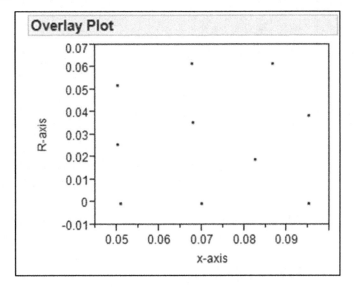

13. Open Space-Filling.jmp. This data table holds the information presented in Table 11.16.

14. From the red triangle next to the **Model**, click **Run Script**. This automatically fits the Gaussian Process Model because the table was formed from the Space Filling Design platform in JMP. The actual versus jackknife predicted plot from Table 11.17 is shown below.

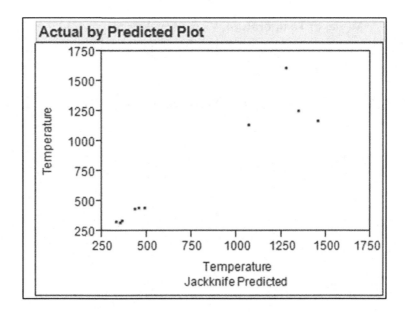

15. To produce the Contour Profiler shown on page 529, click the red triangle next to Gaussian Process Model of Temperature and select **Surface Profiler**.

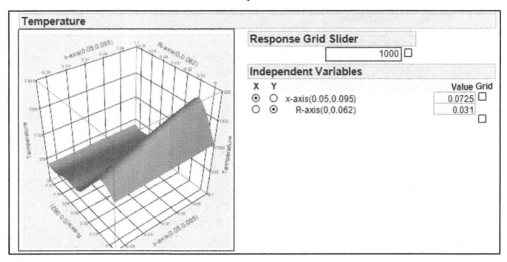

The plot may be rotated to match the view in the book by dragging on the surface.

16. The complex equation shown in Table 11.18 of the textbook reflects the estimates from the Gaussian Process Model. To see your parameter estimates, first select **Save Prediction Formula** from the red triangle next to **Gaussian Process**

Model of Temperature. View the values by going to the data table and clicking the plus sign next to Temperature Prediction Formula in the columns panel.

17. Select **Window > Close All**.

Example 11.5 A Three-Component Mixture

1. Select **DOE > Mixture Design**. Leave the defaults.

Responses

Response Name	Goal	Lower Limit	Upper Limit	Importance
Y	Maximize			

Factors

Name	Role	Values	
X1	Mixture	0	1
X2	Mixture	0	1
X3	Mixture	0	1

2. Click **Continue**.

Choose Mixture Design Type

| Optimal | Create a design tailored to meet specific requirements. |

| Simplex Centroid | Run each ingredient without mixing, then mix equal proportions of K ingredients at a time to the specified limit. | **K** 2 |

| Simplex Lattice | Triangular grid. Specify number of levels per factor: | **Number of Levels** 2 |

| Extreme Vertices | Find the vertices of the simplex. Then add the mid-points of the edges and averages of vertices to the specified degree. | **Degree** 2 |

| Linear Constraint | Add linear constraints on the relative proportions of ingredients. Click once for each constraint. |

| ABCD Design | A mixture design for factor screening. |

Back

3. Next to Simplex Lattice, set the **Number of Levels** to 2.

4. Click **Simplex Lattice**. Note: the ABCD design adds axial runs to the simplex lattice.

5. To match the design of the textbook, set **Number of Replicates** to 2. One of the three replicates of each of the pure blends may then be deleted (at random) to obtain the design of Table 11.19 in the textbook.

6. Click **Make Table**.

7. Open Yarn-Elongation.jmp. The observed elongation values from Table 11.19 have been loaded into this data table. Since the table was created by the Mixture Design platform, the columns have already been coded appropriately for the analysis of a mixture design.

8. From the red triangle next to the **Model**, click **Run Script**. The model specified by this script may be obtained manually by selecting the factors *A*, *B*, and *C* and selecting **Macros > Mixture Response Surface**.

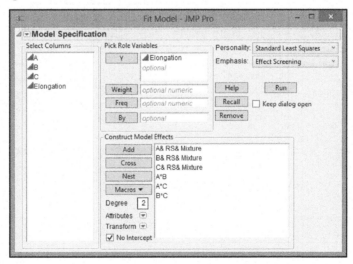

9. Click **Run**.

Parameter Estimates

Term	Estimate	Std Error	t Ratio	Prob>|t|
A	11.7	0.603692	19.38	<.0001*
B	9.4	0.603692	15.57	<.0001*
C	16.4	0.603692	27.17	<.0001*
A*B	19	2.608249	7.28	<.0001*
A*C	11.4	2.608249	4.37	0.0018*
B*C	-9.6	2.608249	-3.68	0.0051*

The positive interactions *A*B* and *A*C* are examples of synergistic blending effects, whereas the negative interaction *B*C* is an example of an antagonistic blending effect.

10. Click the red triangle next to Response Elongation and select **Factor Profiling > Mixture Profiler**.

11. Click the red triangle next to Mixture Profiler and select **Contour Grid**.

12. Click **OK** to accept the default values.

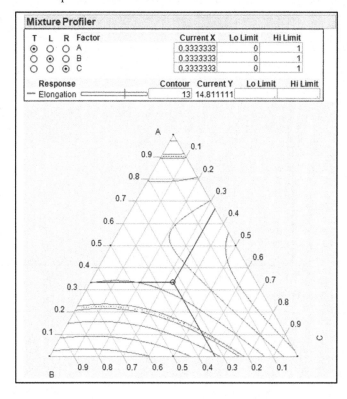

JMP has a special Ternary Plot platform. This plotting feature makes it easy to plot the design points of prospective mixture designs.

13. Click the red triangle next to Response Elongation and select **Save Columns > Prediction Formula**. The prediction formula is now saved in the *Pred Formula Elongation* column in the original data table.

14. Select **Graph > Ternary Plot**.

15. Select *Pred Formula Elongation* for **Contour Formula**.

16. Select *A, B,* and *C* for **X, Plotting**.

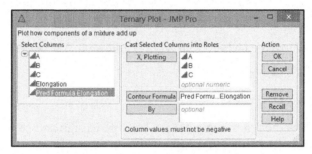

17. Click **OK**.

18. Click the red triangle next to Ternary Plot and select **Contour Fill > Fill Above**.

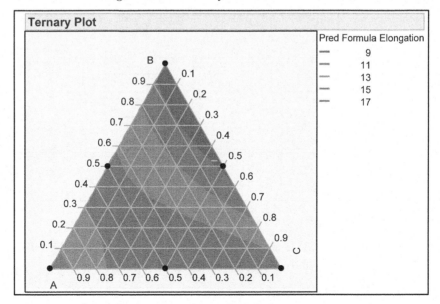

In order to maximize *Elongation*, you need to use a blend of about 70 percent component *C* and 30 percent component *A*. The optimal factor settings may also be found using the prediction profiler that was demonstrated in the previous example.

19. Select **Window > Close All**.

Example 11.6 Paint Formulation

1. Select **DOE > Mixture Design**.

2. Change the response name from *Y* to *Hardness*.

3. Add a second response *Solids* that should be minimized with an upper limit of 30.

4. Change the factors from *X1, X2,* and *X3* to *Monomer, Crosslinker,* and *Resin*.

5. Click **Continue**.

6. Click **Optimal**.

7. Click the red triangle next to Mixture Design and select **Optimality Criterion > Make D-Optimal Design**.

8. Under **Define Factor Constraints**, click **Add Constraint** six times and fill in the spaces to match the constraints shown below.

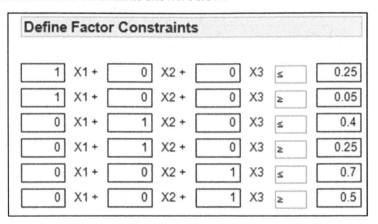

9. In the Model report, select **Interactions > 2nd**.

10. In the Design Generation report, select the option **User Specified** under **Number of Runs**. Request 14 runs.

11. Click **Make Design**.

12. Click **Make Table**.

13. Select **Graph > Ternary Plot**.

14. Select *Monomer, Crosslinker,* and *Resin* for **X, Plotting**.

15. Click **OK**.

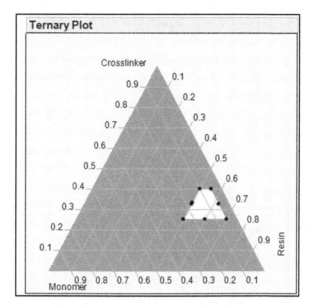

Notice how the runs are distributed throughout the constrained design space. Different optimality criteria will distribute the runs in different ways.

16. Open Paint-Formulation.jmp

17. From the red triangle next to the **Model**, click **Run Script**.

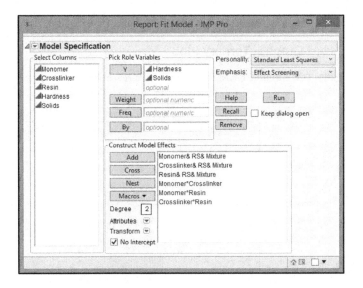

18. Click **Run**.

Response Hardness

Parameter Estimates

Term	Estimate	Std Error	t Ratio	Prob>|t|
(Monomer-0.05)/0.2	23.808867	3.358421	7.09	0.0001*
(Crosslinker-0.25)/0.2	16.397793	7.685953	2.13	0.0654
(Resin-0.5)/0.2	29.450347	3.357532	8.77	<.0001*
Monomer*Crosslinker	44.424842	25.31825	1.75	0.1174
Monomer*Resin	-44.00586	15.94531	-2.76	0.0247*
Crosslinker*Resin	13.797644	23.31954	0.59	0.5704

Response Solids

Parameter Estimates

Term	Estimate	Std Error	t Ratio	Prob>|t|
(Monomer-0.05)/0.2	26.531171	3.993862	6.64	0.0002*
(Crosslinker-0.25)/0.2	46.607193	9.1402	5.10	0.0009*
(Resin-0.5)/0.2	73.224206	3.992805	18.34	<.0001*
Monomer*Crosslinker	-75.7789	30.10868	-2.52	0.0360*
Monomer*Resin	-55.49169	18.96229	-2.93	0.0191*
Crosslinker*Resin	-154.6355	27.73179	-5.58	0.0005*

The *Crossliner*Resin* and *Monomer*Crosslinker* blending effects are significant in the model for *Solids* but not in the model for *Hardness*. Furthermore, the *Crosslinker* main effect has a p-value of 0.0654 in the *Hardness* model, but is strongly significant in the *Solids* model with a p-value of 0.009.

19. Click the red triangle next to Least Squares Fit and select **Profilers > Mixture Profiler**.

20. For *Hardness*, in **Lo Limit**, enter 25. In **Hi Limit**, enter 40.

21. For *Solids*, in **Lo Limit**, enter 0. In **Hi Limit**, enter 30.

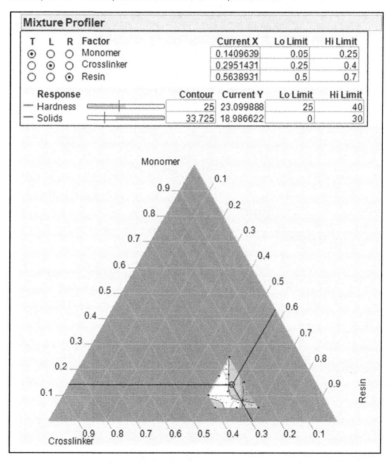

The non-gray trapezoidal region indicates which mixture combinations are allowed by the restrictions on *Monomer, Crosslinker,* and *Resin*. The gray-shaded regions that are outside of the trapezoid designate mixture combinations that are disallowed by restrictions on the individual factors. Finally, the red and blue-shaded regions inside the trapezoid display combinations that produce out-of-limit responses. The white regions within the trapezoid represent ideal factor combinations that satisfy the restrictions on all of the mixture components and the predicted responses. The crosshair may be moved by dragging it, or by setting the values of **Current X.** The prediction profiler introduced in Section 11.3.4 may also be used to find the optimal mixture component settings that match the target values for *Hardness* and *Solids*.

Robust Parameter Design and Process Robustness Studies

The response surface methods introduced in Chapter 11 provide the ability to optimize factor settings in order to produce a mean response that falls within specified limits. These tools may also be used to select factor levels that minimize the variation in the response due to factors that cannot be easily controlled during normal operation of the process. When you control these noise factors during the execution of a process robustness study, you can determine how the inevitable fluctuations in these factors (in the normal operation of the process) will affect the response at different settings of the factors that can be easily controlled.

Robust parameter designs often provide the ability to fit a second-order model in the controllable factors. In addition, the first-order effects of noise factors are modeled, along with their interactions with the controllable factors. The interactions between the controllable and the noise factors are essential. The parameter estimates from this combined array are then used to build two new models: the mean and the variance of the response are modeled as functions of the controllable factors. Although the noise factors do not explicitly appear in either of these models (see point 1 on page 561 of the textbook), the parameter estimates for the control*noise interactions determine the curvature of the resulting surface for the response variance.

The response surfaces for the response mean and variance may then be analyzed by overlaying contour plots or by setting desirability functions. The experimenter may then find the levels of the controllable factors that keep the mean response within the desired range while at the same time minimizing process variability.

Example 12.1 Two Controllable Variables and One Noise Variable

1. Open Pilot-Plant-Filtration.jmp.

2. From the red triangle next to **Model**, click **Run Script**.

3. Remove all but the *Temperature, Conc., Stir Rate, Temperature*Conc.,* and *Temperature*Stir Rate* effects from the Construct Model Effects area.

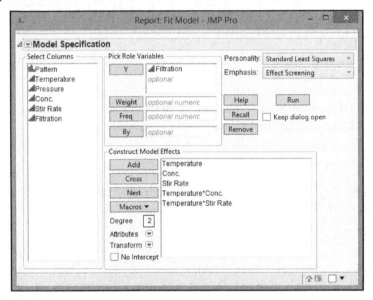

4. Click **Run**.

Analysis of Variance

Source	DF	Sum of Squares	Mean Square	F Ratio
Model	5	5535.8125	1107.16	56.7412
Error	10	195.1250	19.51	Prob > F
C. Total	15	5730.9375		<.0001*

Parameter Estimates

Term	Estimate	Std Error	t Ratio	Prob>|t|
Intercept	70.0625	1.104324	63.44	<.0001*
Temperature	10.8125	1.104324	9.79	<.0001*
Conc.	4.9375	1.104324	4.47	0.0012*
Stir Rate	7.3125	1.104324	6.62	<.0001*
Temperature*Conc.	-9.0625	1.104324	-8.21	<.0001*
Temperature*Stir Rate	8.3125	1.104324	7.53	<.0001*

The models for the mean and variance are derived from these parameter estimates as shown in Example 12.1 in the textbook. All of the parameter estimates involving the noise factor *Temperature* appear only in the model for the variance.

5. Click the red triangle next to Response Filtration and select **Factor Profiling > Contour Profiler**.

6. Set *Conc.* to **Vert** and *Stir Rate* to **Horiz**.

7. Click the red triangle next to **Contour Profiler** and select **Contour Grid**.

8. Fill out the dialog box as shown in the screenshot.

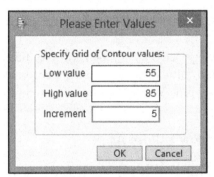

9. Click **OK**.

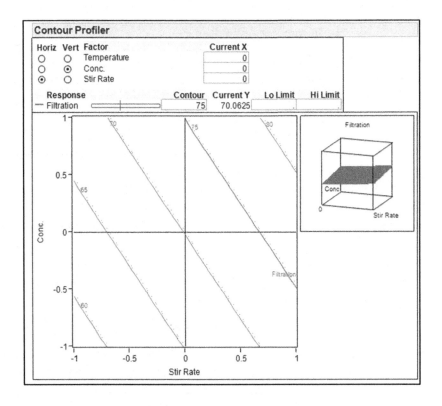

In the model for the process mean, the values for the noise factors (e.g., *Temperature*) are set to their expected value of 0. JMP does not automatically produce the propagation of error (POE) contour plot. We will see how it can be created manually in the next example.

10. Select **Window > Close All**.

Example 12.2 Two Controllable Variables and Three Noise Variables

1. Open Semi-Conductor.jmp.

2. From the red triangle next to **Model**, click **Run Script**.

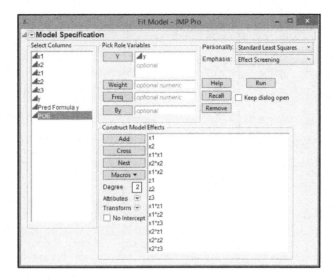

3.   Click **Run**.

Analysis of Variance				
Source	DF	Sum of Squares	Mean Square	F Ratio
Model	14	1456.9884	104.071	109.1319
Error	8	7.6290	0.954	Prob > F
C. Total	22	1464.6174		<.0001*

Parameter Estimates

Term	Estimate	Std Error	t Ratio	Prob>|t|
Intercept	30.365	0.43672	69.53	<.0001*
x1	-2.920833	0.199335	-14.65	<.0001*
x2	-4.129167	0.199335	-20.71	<.0001*
x1*x1	2.5959375	0.221746	11.71	<.0001*
x2*x2	2.1834375	0.221746	9.85	<.0001*
x1*x2	2.86875	0.244134	11.75	<.0001*
z1	2.73125	0.244134	11.19	<.0001*
z2	-2.33125	0.244134	-9.55	<.0001*
z3	2.33125	0.244134	9.55	<.0001*
x1*z1	-0.26875	0.244134	-1.10	0.3030
x1*z2	0.89375	0.244134	3.66	0.0064*
x1*z3	2.58125	0.244134	10.57	<.0001*
x2*z1	2.00625	0.244134	8.22	<.0001*
x2*z2	-1.43125	0.244134	-5.86	0.0004*
x2*z3	1.55625	0.244134	6.37	0.0002*

These parameter estimates are used to build the models for the mean and the variance as shown on page 566 of the textbook.

4. Click the red triangle next to **Response y** and select **Save Columns > Prediction Formula.**

5. Return to the data table, and select the plus sign next to *Pred Formula y* in the columns panel.

6. Delete all of the terms that contain *z1, z2,* or *z3* from the existing formula.

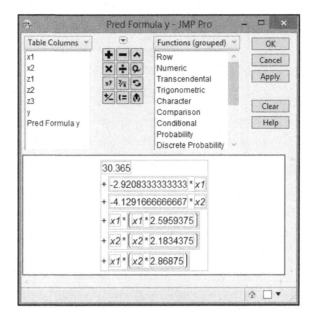

7. Click **OK**.

8. Choose **Cols > New Column**.

9. Enter *POE* for **Column Name**.

10. Select **Formula** from the **Column Properties** drop-down menu.

11. Fill out the formula for the POE as it appears in the screenshot. For convenience, this column has been included with the Semi-Conductor data table.

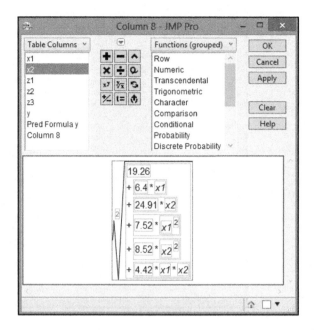

12. Click **OK** twice.

13. Select **Graph > Contour Profiler**.

14. Select *Pred Formula y* and *POE* for **Y, Prediction Formula**.

15. Click **OK**.

16. Enter 30 in **Hi Limit** for **Pred Formula y**.

17. Enter 5 in **Hi Limit** for POE.

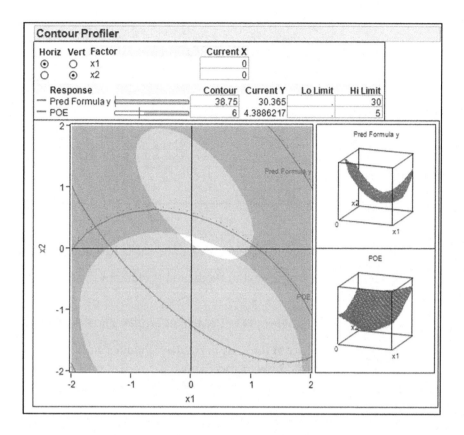

The narrow band of white space in the middle of the plot is the acceptable region for the settings of the controllable factors. The red and blue bands indicate which areas are within the target range for *Filtration* and *POE*, respectively.

18. The Contour Profiler platform provides a similar functionality for minimizing the impact of noise factors. Return to the Semi-Conductor – Fit Least Squares window (the Fit Model platform).

19. Click the red triangle next to Response y and select **Save Columns > Prediction Formula.** A new column *Pred Formula y 2* will appear in the data table. The formula in this column depends on the noise factors that were deleted from the formula appearing in the *Pred Formula y* column.

20. Select **Graph > Contour Profiler**.

21. Select *Pred Formula y 2* for **Y, Prediction Formula**.

22. Select *z1, z2,* and *z3* for **Noise Factors.**

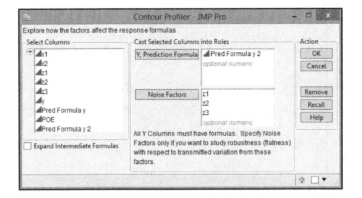

23. Click **OK**.

24. Click the red triangle next to Profiler and select **Prediction Profiler**.

25. Click the red triangle next to Prediction Profiler and select **Desirability Functions.**

26. Click the red triangle next to Prediction Profiler and select **Set Desirabilities.**

27. Select **Match Target** for Pred Formula y 2 Values and enter 30, 29, and 27 for **High, Middle,** and **Low,** respectively.

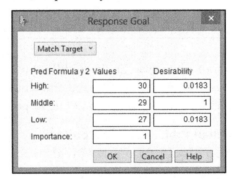

28. Click **OK**.

29. Click **OK** three more times, matching the derivatives of the prediction formula with respect to the noise factors to a target of 0.

30. Click the red triangle next to Prediction Profiler and select **Maximize Desirability**.

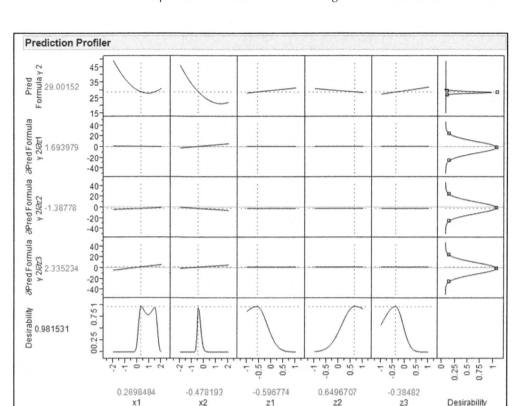

The prediction profiler recommends setting $x1 = 0.2898$ and $x2 = -0.4782$ to obtain a predicted y of 29.

31. Return to the Contour Profiler in the window titled "Semi-Conductor – Contour Profiler of Pred Formula y, POE."

32. Enter 0.2898 for the $x1$ value of **Current X**.

33. Enter -0.4782 for the $x2$ value of **Current X**.

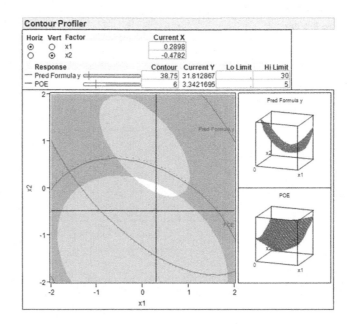

The factor settings recommended by the prediction profiler appear at the crosshairs on the Contour Profiler. This point falls outside of the white region recommended by the transmission of error approach using propagation of error (POE).

34. Select **Window > Close All**.

Experiments with Random Factors

For some experiments, the factor levels included in the design are the only ones of interest. For example, a grocery store may be considering one of four possible layouts for a soft-drink display. A randomized experiment will allow the store to determine if there is a difference in sales across the four layouts, and the layout with the best sales may be chosen. By contrast, some factor levels are only samples of a large population of possibilities. A manufacturer may want to test for operator-to-operator variability in a process. If the manufacturer intends to draw inference about the entire population of operators, not just those limited number of operators included in the experiment, the operator factor should be treated as a random effect. Models with only random factors are referred to as random effects models, while those with both random and fixed factors are known as mixed models. When a factor is treated as a fixed effect, the test for significance investigates whether or not the mean contribution of the effect is different from 0. For a random effect, the test focuses on whether the variance of the (normally distributed) factor population is greater than 0. This is referred to as the variance component of the factor.

JMP supports two different methods for estimating mixed models. The first is the method of moments, expected mean squares (EMS) method. This is the primary method presented in the textbook. Just as with the fixed effects models, F-tests are constructed to test for the significance of each factor's contribution. However, when compared to fixed effect models, the ANOVA calculations change in the presence of random effects. The appropriate ratios of mean squares can change. JMP automatically calculates and reports the appropriate ratios. According to the JMP documentation, "For historical interest only, the [Fit Model] platform also offers the Method of Moments (EMS), but this is no longer a recommended method… the reason to prefer REML is that it works without depending on balanced data, or shortcut approximations, and it gets all the tests right, even contrasts that work across interactions. Most packages that use the traditional EMS method are either not able to test some of these contrasts, or compute incorrect variances for them."

JMP furnishes the ability to obtain restricted maximum likelihood (REML) estimates. These estimates are the same as those that would be obtained from PROC MIXED in SAS. This approach does not require the calculation of expected mean squares or approximate F-tests. REML is an asymptotic method that relies on the central limit theorem. Care should be taken when using the 95% confidence intervals for the variance components that are produced by JMP when the "Unbounded Variance Components" option is checked. These intervals, commonly known as Wald intervals, are known to be too wide when the variance component is actually 0 or when the sample size is small. That is, there will be many times when a variance component is significantly greater than 0 (at the 5% significance level), but the 95% Wald confidence interval will contain 0. If the "Unbounded Variance Components" option is not selected, then Satterthwaite confidence intervals are produced for the variance components. These confidence intervals are based on a chi-square distribution and thus asymmetric. These intervals are more appropriate than the Wald intervals for small sample sizes or when the true variance component is near 0. The Satterthwaite intervals will not contain 0 unless the solution for the variance component is equal to 0: the lower bounds of these confidence intervals may be checked for practical significance, and compared to the magnitudes of other variance components in the model. A modified likelihood-ratio test provides the most rigorous technique for testing whether variance components are significantly greater than 0, though a description of this test is beyond the scope of this book.

Example 13.1 A Measurement Systems Capability Study

1. Open Measurement-System.jmp.

2. Select **Analyze > Fit Model**.

3. Select *y* for **Y**.

4. Select *Parts* and *Operators* and select **Macros > Full Factorial**.

5. Select the terms *Parts, Operators,* and *Parts*Operators* in the Construct Model Effects area.

6. Click the red triangle next to **Attributes** and select **Random Effect**.

7. To match the output in the textbook, click the button next to **Method** and select **EMS (Traditional)**.

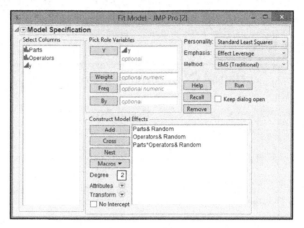

8. Click **Run**.

Variance Component Estimates

Component	Var Comp Est	Percent of Total
Parts&Random	10.27982	92.225
Operators&Random	0.014912	0.134
Parts*Operators&Random	-0.13991	-1.255
Residual	0.991667	8.897
Total	11.14649	100.000

These estimates based on equating Mean Squares to Expected Value.

Tests wrt Random Effects

Source	SS	MS Num	DF Num	F Ratio	Prob > F
Parts&Random	1185.43	62.3908	19	87.6470	<.0001*
Operators&Random	2.61667	1.30833	2	1.8380	0.1730
Parts*Operators&Random	27.05	0.71184	38	0.7178	0.8614

The *Parts*Operators* interaction is not significant and may be removed.

9. Click the red triangle next to Response *y* and select **Model Dialog**.

10. Click the **Random Effects** tab in the Construct Model Effects section.

11. Select *Parts*Operators& Random* under Construct Model Effects and click **Remove**.

12. Click **Run**.

Variance Component Estimates

Component	Var Comp Est	Percent of Total
Parts&Random	10.25127	91.980
Operators&Random	0.010629	0.095
Residual	0.883163	7.924
Total	11.14506	100.000

These estimates based on equating Mean Squares to Expected Value.

Tests wrt Random Effects

Source	SS	MS Num	DF Num	F Ratio	Prob > F
Parts&Random	1185.43	62.3908	19	70.6447	<.0001*
Operators&Random	2.61667	1.30833	2	1.4814	0.2324

The part-to-part variation contributes significantly to the overall variation, whereas there is no evidence of significant operator-to-operator variation. The part-to-part variation contributes to over 90% of the total process variation.

13. Close the analysis reports but leave the Measurement-System data table open for the next exercise.

Example 13.3 The Unrestricted Model

Note that we have skipped Example 13.2. JMP does not support the restricted model.

1. Return to the Measurement-System data table. We will perform the same analysis as in Example 13.1 using REML instead of EMS.

2. Choose **Analyze > Fit Model**.

3. Select *Parts* and *Operators* and click **Macros > Full Factorial**.

4. Select *y* for **Y**.

5. Select the terms *Parts* and *Parts*Operators* in the Construct Model Effects area, holding the *Ctrl* key while clicking each effect.

6. Click the red triangle next to **Attributes** and select **Random Effect**.

7. Check **Keep dialog open**.

8. Click **Run**.

REML Variance Component Estimates

Random Effect	Var Ratio	Var Component	Std Error	95% Lower	95% Upper	Pct of Total
Parts	10.36621	10.279825	3.3738173	3.6672642	16.892385	91.202
Parts*Operators	-0.141088	-0.139912	0.1219114	-0.378854	0.0990296	0.000
Residual		0.9916667	0.1810527	0.7143057	1.4697982	8.798
Total		11.271491	3.3786718	6.7595239	22.472214	100.000

-2 LogLikelihood = 410.4121524

Note: Total is the sum of the positive variance components.

Total including negative estimates = 11.131579

Fixed Effect Tests

Source	Nparm	DF	DFDen	F Ratio	Prob > F
Operators	2	2	38	1.8380	0.1730

This output matches Table 13.6 in the textbook.

9. Return to the **Fit Model** dialog.

10. Deselect **Unbounded Variance Components**.

11. Click **Run**.

REML Variance Component Estimates

Random Effect	Var Ratio	Var Component	Std Error	95% Lower	95% Upper	Pct of Total
Parts	11.607447	10.251271	3.373773	5.8888058	22.154905	92.068
Parts*Operators	0	0	0	0	0	0.000
Residual		0.8831633	0.1261662	0.6799858	1.1937776	7.932
Total		11.134434	3.3753454	6.6438537	22.403646	100.000

-2 LogLikelihood = 411.65438594

Note: Total is the sum of the positive variance components.

Total including negative estimates = 11.134434

Fixed Effect Tests

Source	Nparm	DF	DFDen	F Ratio	Prob > F
Operators	2	2	98	1.4814	0.2324

In the second analysis presented here (with the **Unbounded Variance Components** option not selected), the variance component for the *Parts*Operators* effect is reported to be 0. In this case, the Fit Model platform removes the *Parts*Operators* effect from the analysis. The reported values for the remaining parameters are identical to what they would have been if *Parts*Operators* had been omitted from the Construct Model Effects area. Finally, notice how the confidence intervals for the variance components in the second analysis are wider and right-skewed since they are based on a chi-squared instead of a normal distribution.

12. Select **Window > Close All**.

Example 13.5 A Three-Factor Factorial Experiment with Random Factors

1. Open Turbine-Experiment.jmp.

2. Select **Analyze > Fit Model**.

3. Select *Drop* for **Y**.

4. Select *Gas Temp, Operator,* and *Pressure Gauge*.

5. Select **Macros > Full Factorial**.

6. Select all of the terms that now appear in the Construct Model Effects area.

7. Click the red triangle next to **Attributes** and select **Random Effect**.

8. To match the output of the textbook, select **EMS (Traditional)** from the **Method** drop-down menu.

9. Check **Keep dialog open**.

10. Click **Run**.

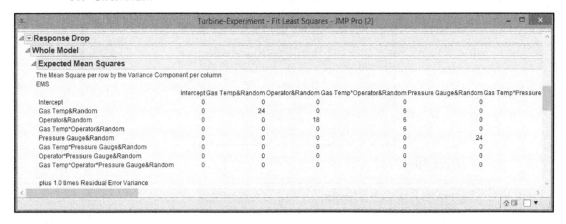

JMP produces a table that provides the expected mean squares for each term.

11. Leave the windows from this analysis open for the next example.

Example 13.6 Approximate F Tests

1. Return to the **Fit Model** dialog.

2. Select *Gas Temp& Random* in the Construct Model Effects area.

3. Click the red triangle next to **Attributes** and deselect **Random Effect**.

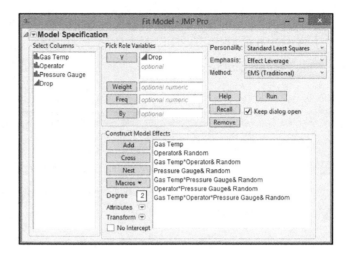

4. Click **Run**.

Tests wrt Random Effects

Source	SS	MS Num	DF Num	F Ratio	Prob > F
Gas Temp	1023.36	511.681	2	2.2984	0.1712
Operator&Random	423.819	141.273	3	0.6333	0.6164
Gas Temp*Operator&Random	1211.97	201.995	6	14.5923	<.0001*
Pressure Gauge&Random	7.19444	3.59722	2	0.0648	0.9379
Gas Temp*Pressure Gauge&Random	137.889	34.4722	4	2.4903	0.0991
Operator*Pressure Gauge&Random	209.472	34.912	6	2.5221	0.0814
Gas Temp*Operator*Pressure Gauge&Random	166.111	13.8426	12	0.6468	0.7882

5. Click the red triangle next to Response Drop and select **Factor Profiling > Interaction Plots**.

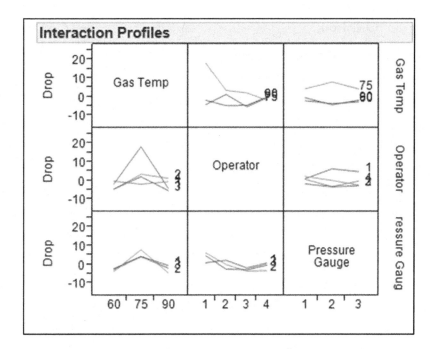

There is strong visual evidence of a *Gas Temp*Operator* interaction effect.

6. Return to the **Fit Model** dialog.

7. For the **Method** option, select **REML (Recommended).**

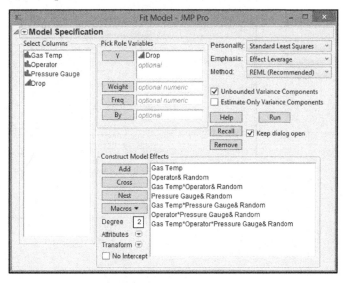

8. Click **Run**.

REML Variance Component Estimates

Random Effect	Var Ratio	Var Component	Std Error	95% Lower	95% Upper	Pct of Total
Operator	-0.212308	-4.543981	9.1867341	-22.54965	13.461687	0.000
Gas Temp*Operator	1.4651741	31.358796	19.459821	-6.781751	69.499344	53.284
Pressure Gauge	-0.101125	-2.164352	1.3471464	-4.80471	0.4760065	0.000
Gas Temp*Pressure Gauge	0.1204845	2.5787037	3.127757	-3.551587	8.7089947	4.382
Operator*Pressure Gauge	0.1640709	3.5115741	3.4889495	-3.326641	10.349789	5.967
Gas Temp*Operator*Pressure Gauge	-0.176617	-3.780093	3.7876462	-11.20374	3.6435576	0.000
Residual		21.402778	5.0446831	14.1539	36.112874	36.367
Total		58.851852	20.749088	32.689863	135.92987	100.000

-2 LogLikelihood = 434.10413622
Note: Total is the sum of the positive variance components.
Total including negative estimates = 48.363426

Fixed Effect Tests

Source	Nparm	DF	DFDen	F Ratio	Prob > F
Gas Temp	2	2	6.967	2.2984	0.1712

9. Return to the **Fit Model** dialog.

10. Deselect the **Unbounded Variance Components** option.

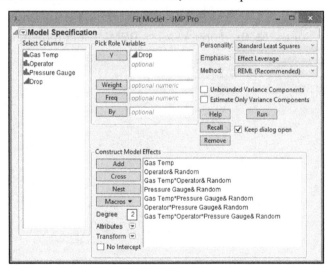

11. Click **Run**.

REML Variance Component Estimates

Random Effect	Var Ratio	Var Component	Std Error	95% Lower	95% Upper	Pct of Total
Operator	0	0	0	0	0	0.000
Gas Temp*Operator	1.3376999	26.743799	14.337854	11.668333	111.41745	54.903
Pressure Gauge	0	0	0	0	0	0.000
Gas Temp*Pressure Gauge	0.0353765	0.7072597	2.0000713	0.0779342	9.269e+11	1.452
Operator*Pressure Gauge	0.0634154	1.2678238	2.4449491	0.1906466	452853.12	2.603
Gas Temp*Operator*Pressure Gauge	0	0	0	0	0	0.000
Residual		19.992376	4.2168696	13.751759	31.72403	41.043
Total		48.711259	14.77983	29.054019	98.084581	100.000

-2 LogLikelihood = 438.09850961
Note: Total is the sum of the positive variance components.
Total including negative estimates = 48.711259

Fixed Effect Tests

Source	Nparm	DF	DFDen	F Ratio	Prob > F
Gas Temp	2	2	9.082	2.7493	0.1165

Notice that the *Gas Temp*Operator* and *Residual* (error) components are the only ones to have a lower 95% Satterthwaite confidence bond greater than 0.2 in the final analysis. Together, these components account for over 95% of the total process variation. This output suggests that the *Gas Temp*Operator* effect is important. Contrast this with the 95% Wald confidence interval for the same effect (-6.78 to 69.50). This interval contains 0 and would wrongly lead to the conclusion that the effect is not important.

The fixed effect tests for *Gas Temp* in the last two analyses differ because the *Operator, Pressure Gauge,* and *Gas Temp*Operator*Pressure Gauge* variance components are reported to be 0 (and thus dropped from the model) when **Unbounded Variance Components** is not selected. The last displayed Fixed Effect Tests report is identical to the one that would be produced if the **Unbounded Variance Components** option were selected and the *Operator, Pressure Gauge,* and *Gas Temp*Operator*Pressure Gauge* effects were dropped from the model.

12. Select **Window > Close All**.

Nested and Split-Plot Designs

In the full factorial designs that have been covered in prior chapters, distinct factors A and B in the experiment are crossed. That is, each level of A appears with each level of B in at least one experimental run. However, in some situations it is not possible for all of the combinations of factor levels to appear in a design. For example, if a manufacturer designs an experiment to test the variability in purity of a raw material, two potential factors of interest are the suppliers and the batches from each supplier. In this case, Batch 1 from Supplier 1 is different from Batch 1 from Supplier 2: the batches are nested within the suppliers. As a result, it is not possible to calculate an interaction between Batch and Suppliers.

In many cases, the higher-level factor (e.g., Supplier) is fixed and the lower-level factor (e.g., Batch) is random. The methods discussed in Chapter 13 are then used to estimate the variance components for the random factors.

It is also possible to have more than two layers of nesting. JMP will allow up to ten levels of nesting. It is important to keep the design structure in mind when specifying a model to analyze the experimental results. No nested factors may be involved in an interaction with any of the factors under which they were nested.

Split-plot designs provide a form of randomization restriction in experimental designs. This restriction allows hard-to-change factors to remain constant until all runs at each level of that factor have been completed. The hard to change factors are organized into whole plots, and the easy-to-change factors are organized into subplots. The whole plot factor (or factor combinations) are randomized *between* each whole plot; the subplot factors are randomized *within* each whole plot. The restriction leads to two different randomization schemes. These schemes in turn lead to the calculation of two different error terms: one for the whole plot factors and another for the subplot factors. If possible, it is best to assign the most interesting factors to subplots, since any uncontrolled or undesigned factors that vary with the whole plot factors will be completely confounded with the whole plot effects.

Example 14.1 The Two-Stage Nested Design

1. Open Purity-Data.jmp.

2. Select **Analyze > Fit Model**.

3. Select *Y* for **Y**.

4. Select both *Batch* and *Supplier* and click **Add**.

5. Select *Supplier* under Select Columns, and select *Batch* under Construct Model Effects.

6. Click **Nest**.

7. Select the *Batch[Supplier]* effect in the Construct Model Effects area.

8. Click the red triangle next to **Attributes** and select **Random Effect**.

9. Check **Keep dialog open.**

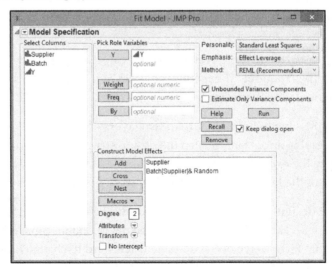

10. Click **Run**.

Summary of Fit

RSquare	0.518555
RSquare Adj	0.489376
Root Mean Square Error	1.624466
Mean of Response	0.361111
Observations (or Sum Wgts)	36

REML Variance Component Estimates

Random Effect	Var Ratio	Var Component	Std Error	95% Lower	95% Upper	Pct of Total
Batch[Supplier]	0.6479532	1.7098765	1.2468358	-0.733877	4.1536298	39.319
Residual		2.6388889	0.7617816	1.6089119	5.1070532	60.681
Total		4.3487654	1.3221333	2.5915066	8.7710941	100.000

-2 LogLikelihood = 145.04119391

Note: Total is the sum of the positive variance components.

Total including negative estimates = 4.3487654

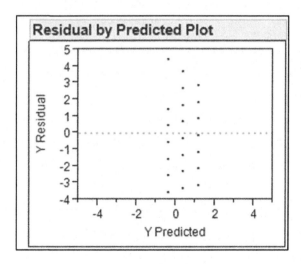

This output matches Table 14.7 in the textbook. From the Residual by Predicted plot, it appears that the batch-to-batch variability is about the same for each supplier. The *Batch* variance component accounts for around 40% of the overall process variability. Even though the 95% confidence interval for this effect contains 0, we should not discard it. See Chapter 13 for further discussion about confidence intervals for variance components.

11. Return to the **Fit Model** dialog.

12. Deselect the option for **Unbounded Variance Components**.

☐ Unbounded Variance Components

Since none of the estimated variance components were negative, the resulting parameter estimates will be the same as they were with the option selected. (When some variances components are constrained to 0, they are dropped from the analysis. The resulting tests for fixed effects in the model may be affected.) However, JMP will now produce Satterthwaite confidence intervals for the variance components, which are more accurate than the Wald intervals. Again, see Chapter 13 for further discussion.

13. Click **Run**.

REML Variance Component Estimates

Random Effect	Var Ratio	Var Component	Std Error	95% Lower	95% Upper	Pct of Total
Batch[Supplier]	0.6479532	1.7098765	1.2468358	0.5995954	15.59716	39.319
Residual		2.6388889	0.7617816	1.6089119	5.1070532	60.681
Total		4.3487654	1.3221333	2.5915066	8.7710941	100.000

-2 LogLikelihood = 145.04119391
Note: Total is the sum of the positive variance components.
Total including negative estimates = 4.3487654

Fixed Effect Tests

Source	Nparm	DF	DFDen	F Ratio	Prob > F
Supplier	2	2	9	0.9690	0.4158

The Satterthwaite confidence intervals will not contain 0 unless a variance component is estimated to be 0. Instead, we notice that even the lower bound of the confidence interval of 0.60 still accounts for a sizable portion of the overall variation.

14. Select **Window > Close All**.

Example 14.2 A Nested-Factorial Design

1. Open Assembly-Time.jmp.

2. Select **Analyze > Fit Model**.

3. Select *Time* for **Y**.

4. Select *Layout, Operator,* and *Fixture.*

5. Select **Macros > Full Factorial**.

6. Select *Layout*Operator* and click **Remove**.

7. Answer **Yes** to the prompt "Remove other effects containing selected effect in the model?"

8. Select *Layout* under Select Columns.

9. Select *Operator* and *Operator*Fixture* under Construct Model Effects.

10. Click **Nest**.

11. Click the red triangle next to **Attributes** and select **Random Effect**.

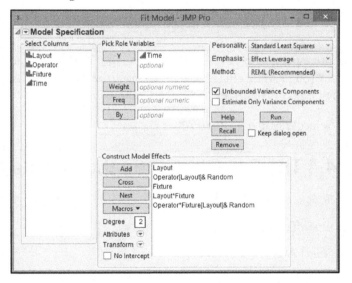

12. Click **Run**.

REML Variance Component Estimates

Random Effect	Var Ratio	Var Component	Std Error	95% Lower	95% Upper	Pct of Total
Operator[Layout]	0.4642857	1.0833333	1.2122659	-1.292664	3.4593308	21.697
Operator*Fixture[Layout]	0.6755952	1.5763889	1.1693951	-0.715583	3.8683612	31.572
Residual		2.3333333	0.6735753	1.4226169	4.5157102	46.732
Total		4.9930556	1.4145791	3.0688692	9.5261827	100.000

-2 LogLikelihood = 195.88509411
Note: Total is the sum of the positive variance components.
Total including negative estimates = 4.9930556

Fixed Effect Tests

Source	Nparm	DF	DFDen	F Ratio	Prob > F
Layout	1	1	6	0.3407	0.5807
Fixture	2	2	12	7.5456	0.0076*
Layout*Fixture	2	2	12	1.7354	0.2178

This output agrees with Table 14.15 of the textbook. The fixed effects tests indicate that there is a strong *Fixture* effect. *Operators[Layout]* and *Operators*Fixture[Layout]* each make relatively large contributions to the total process variation.

13. Select **Window > Close All**.

Section 14.4 The Experiment on the Tensile Strength of Paper

1. Open Tensile-Strength.jmp.

2. Select **Analyze > Fit Model**.

3. Select *Strength* for **Y**.

4. Select *Block, Pulp,* and *Temperature*.

5. Ensure **Degree** is listed as 2. Select **Macros > Factorial to Degree**.

6. Delete the *Block*Temperature* effect.

7. Select *Block* and *Block*Pulp* under Construct Model Effects, holding down the *Ctrl* key as you click each effect.

8. Click the red triangle next to **Attributes** and select **Random Effect**.

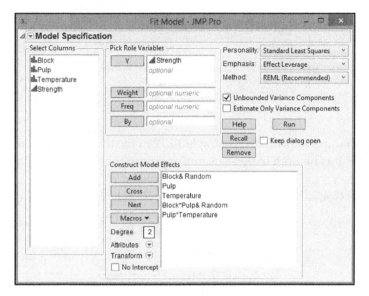

9. Click **Run**.

Summary of Fit

RSquare	0.903675
RSquare Adj	0.859526
Root Mean Square Error	1.993043
Mean of Response	36.02778
Observations (or Sum Wgts)	36

REML Variance Component Estimates

Random Effect	Var Ratio	Var Component	Std Error	95% Lower	95% Upper	Pct of Total
Block	0.6232517	2.4756944	3.2753747	-3.943922	8.8953109	32.059
Block*Pulp	0.3208042	1.2743056	1.6370817	-1.934316	4.4829267	16.502
Residual		3.9722222	1.3240741	2.2679421	8.6869402	51.439
Total		7.7222222	3.5455691	3.7118502	24.719671	100.000

-2 LogLikelihood = 139.36226272
Note: Total is the sum of the positive variance components.
Total including negative estimates = 7.7222222

Fixed Effect Tests

Source	Nparm	DF	DFDen	F Ratio	Prob > F
Pulp	2	2	4	7.0781	0.0485*
Temperature	3	3	18	36.4266	<.0001*
Pulp*Temperature	6	6	18	3.1538	0.0271*

This output matches that of Table 14.19 in the textbook. The *Pulp*Temperature* interaction is significant. Both of these factors are significant in determining the strength of the produced paper. To reproduce the results in 14.20, simply go to **Analyze > Fit Model**, select **Recall**, delete the *Block* and *Block by Pulp* factors, and click **Run**.

Summary of Fit

RSquare	0.7748
RSquare Adj	0.671583
Root Mean Square Error	2.778889
Mean of Response	36.02778
Observations (or Sum Wgts)	36

Analysis of Variance

Source	DF	Sum of Squares	Mean Square	F Ratio
Model	11	637.63889	57.9672	7.5065
Error	24	185.33333	7.7222	Prob > F
C. Total	35	822.97222		<.0001*

Effect Tests

Source	Nparm	DF	Sum of Squares	F Ratio	Prob > F
Pulp	2	2	128.38889	8.3129	0.0018*
Temperature	3	3	434.08333	18.7374	<.0001*
Pulp*Temperature	6	6	75.16667	1.6223	0.1843

This output matches that of Table 14.20 in the textbook. Notice that the *Pulp*Temperature* interaction is not significant when the split-plot design of the experiment is (incorrectly) disregarded.

10. Select **Window > Close All**.

Example 14.3 A 2⁵⁻¹ Split-Plot Experiment

1. Open Wafer-Uniformity.jmp.

2. Select **Analyze > Fit Model**.

3. Select *Uniformity* for **Y**.

4. Select *A, B,* and *C* under Select Columns.

5. Ensure that **Degree** is listed as 2. Select **Macros > Factorial to Degree**.

6. Check **Keep dialog open**.

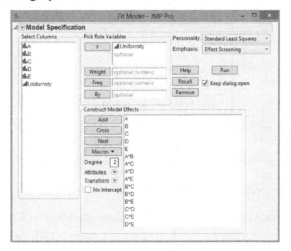

7. Click **Run**.

8. Click the red triangle next to Response Uniformity and select **Effect Screening > Normal Plot**.

9. Click the red triangle next to Normal Plot and select **Half Normal Plot**.

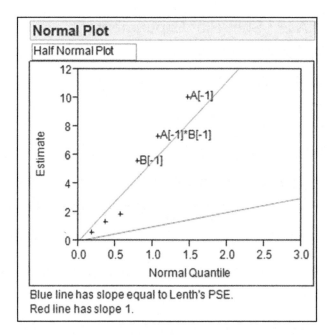

Of the whole plot factors, *A, B,* and *A*B* have large effects.

10. Return to the **Fit Model** dialog.

11. Remove any effects from the Construct Model Effects area that do not include the subplot effects *D* or *E* under the Construct Model Effects. Answer **No** to the prompt "Remove other effects containing selected effect in the model?"

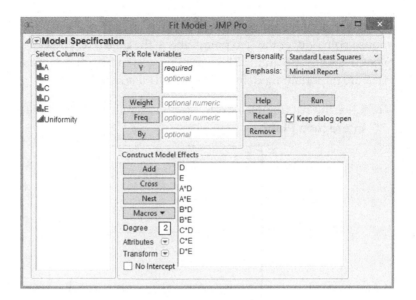

12. Click **Run**.

13. Click **Continue** when prompted that "The model is missing an effect."

14. Click the red triangle next to Response Uniformity and select **Effect Screening > Normal Plot.**

15. Click the red triangle next to Normal Plot and select **Half Normal Plot**.

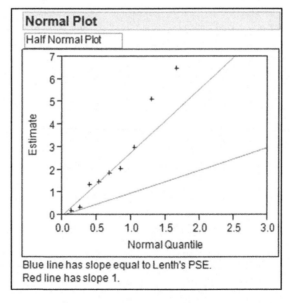

The procedure does not automatically identify any large effects. But when you position the mouse pointer over the two points above the blue line, you can see that the largest subplot effects are $A*E$ and E. The screening design has thus discovered that $A, B, E, A*B,$ and $A*E$ are important modeling factors for *Uniformity*.

16. Click the red triangle next to Response Uniformity and select **Factor Profiling > Interaction Plots**.

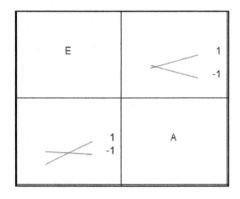

Both levels of E will produce roughly the same *Uniformity* if A is at the low level. If A is at the high level, E should be set to the low level.

17. Select **Window > Close All**.

Other Design and Analysis Topics

This chapter highlights several additional topics related to the design and analysis of experiments. The regression methods covered throughout most of the textbook assume that the response is normally distributed with constant variance. When the observed data do not satisfy these assumptions, we may either transform the response or we may modify the modeling assumptions.

Rather than attempting to find an appropriate transformation via trial and error, the Box-Cox method may be used to select the optimal power transformation of the data. JMP provides this functionality in the Fit Model platform. Transformations are often useful; however, they sometimes lead to nonsensical predictions. For example, it is possible to obtain negative predictions for the square root of a response (e.g., number of defects). Better predictions with smaller confidence intervals often may be obtained by modifying the model itself.

Generalized linear models (GLM) provide flexibility for modeling potentially nonlinear relationships between a linear combination of model terms and the mean response via different link functions. Furthermore, the response can be from any distribution in the exponential family. JMP allows the response distribution to be normal, binomial, Poisson, or exponential. Setting the distribution to binomial and selecting a logit link provides the ability to perform logistic regression, one of the most widespread applications of GLMs. In fact, JMP provides specialized personalities in the Fit Model platform for nominal and logistic regression. Logistic regression allows experimenters to find factors that are significantly associated with a binary response (e.g., whether a product meets specifications).

In earlier chapters, we saw that the blocking principle allows us to reduce experimental error and increase the precision of parameter estimates by removing the effect of controllable nuisance factors. Regression methods allow for modeling of measurable but uncontrollable noise factors. Depending on the number of replicates of the experimental design and on the number of available degrees of freedom, interactions between the nuisance covariates and the design factors may sometimes be calculated.

Example 15.1 Box-Cox Transformation

1. Open Peak-Discharge.jmp.

2. From the red triangle next to **Model**, click **Run Script**.

3. Click **Run**.

4. Click the red triangle next to Response Discharge and select **Factor Profiling > Box Cox Y Transformation**.

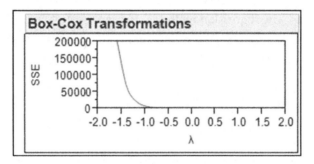

It is necessary to zoom in to find the minimum SSE. You can drag the extreme values of the axes in order to zoom in and out. You can also see the exact Sums of Squares values for common values of λ by clicking the red triangle next to Box-Cox transformations and selecting **Table of Estimates** to see values presented in the text. Alternatively, right-click on the Y axis and select **Axis Settings**. Then set the minimum to 30 and the maximum to 60 and the increment to 10.

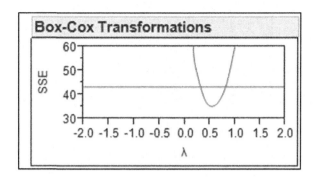

The minimum value occurs near 0.5, suggesting a square root transformation. The transformed response may be saved to the data table by clicking the red triangle next to Box-Cox transformations and selecting **Save Specific Transformation**.

5. Select **Window > Close All**.

Example 15.2 The Generalized Linear Model and Logistic Regression

1. Open Coupon-Redemption.jmp

2. Select **Analyze > Fit Model**.

3. Select *Coupons* and *Events* (in that order) for **Y**.

4. From the drop-down menu for **Personality**, select **Generalized Linear Model**.

5. For **Distribution**, select **Binomial. Logit** is set as the link function by default.

6. Select *A, B,* and *C* under **Select Columns**.

7. Select **Macros > Factorial to Degree**. Ensure that **Degree** is set to 2.

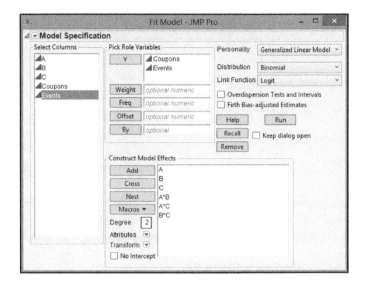

8. Click **Run**.

Parameter Estimates

Term	Estimate	Std Error	L-R ChiSquare	Prob>ChiSq	Lower CL	Upper CL
Intercept	-1.011545	0.025515	1787.942	<.0001*	-1.06176	-0.961737
A	0.1692084	0.0255092	44.269326	<.0001*	0.1192671	0.2192706
B	0.1696223	0.025515	44.470394	<.0001*	0.1196703	0.2196966
C	0.0233173	0.0255099	0.8356425	0.3606	-0.026674	0.0733331
A*B	-0.006285	0.0255122	0.0607035	0.8054	-0.056311	0.0437041
A*C	-0.002773	0.0254324	0.0118847	0.9132	-0.052627	0.0470749
B*C	-0.04102	0.0254339	2.6022317	0.1067	-0.09089	0.0088176

The effects *A* and *B* are significant at the 0.05 level. The two factor interaction term *BC* is marginally significant with a p-value near 0.10. We could now fit a reduced model that does not include the *AB* and *AC* interaction as shown in the example.

Note: It is possible to output the odds ratios shown on page 648 in JMP. The data table must be modified creating a new response variable called *Redeemed* that must be specified as nominal. This variable will have the value of 0 for all 8 runs shown. These 8 runs need to be replicated for a total of 16 runs. The value of *Coupons* for these new 8 runs will be the complement of the original run (that is, 1000 – the original value shown). The newly created *Redeemed* variable will have a value of 1 for all 8 of these runs. Fit the model as described above. Note that the response is now the new nominal variable *Redeemed*, which will default the personality to nominal logistic regression. Ensure the input for the Freq field is *Coupons*. The data table (Coupon-Redemption-Logistic Regression.jmp) is provided for self-study.

9. Select **Window > Close All**.

Example 15.3 Poisson Regression

1. Open Grill.jmp.

2. Select **Analyze > Fit Model**.

3. Select *Defects* for **Y**.

4. Select *A* through *J* under **Select Columns** and click **Add**.

5. From the **Personality** drop-down menu, select **Generalized Linear Model**.

6. Select **Poisson** for Distribution. The canonical **Log** link function is selected by default.

7. Check **Keep dialog open**.

8. Click **Run**.

Parameter Estimates

Term	Estimate	Std Error	L-R ChiSquare	Prob>ChiSq	Lower CL	Upper CL
Intercept	1.1328704	0.1797186	26.426618	<.0001*	0.7538973	1.4619087
A	-0.223604	0.1436625	2.29755	0.1296	-0.504724	0.0690773
B	-0.105947	0.1132135	0.8794856	0.3483	-0.330931	0.115713
C	-0.32471	0.1196879	7.8450841	0.0051*	-0.569329	-0.096094
D	-0.786696	0.1356395	39.641702	<.0001*	-1.071558	-0.533294
E	-0.684839	0.1356395	29.623778	<.0001*	-0.969179	-0.430609
F	-1.095214	0.1436625	91.883951	<.0001*	-1.400052	-0.831958
G	0.0778286	0.1132135	0.4742816	0.4910	-0.143867	0.3027791
H	0.2965913	0.1196879	6.5231748	0.0106*	0.0679402	0.5411834
J	0.0492657	0.1797186	0.0749961	0.7842	-0.309938	0.4051157

9. Click the red triangle next to Generalized Linear Model Fit and select **Save Columns > Predicted Values**.

10. Return to the **Fit Model** dialog.

11. Change **Personality** to **Standard Least Squares**. This will produce an ordinary linear regression of the factors on *Defects*. We could also have changed **Distribution** to **Normal** and **Link** to **Identity** under the **Generalized Linear Model** personality. This method will yield the same parameter estimates. But it will underestimate the standard errors since it obtains the parameter estimates via maximum likelihood estimation.

12. Click **Run**.

13. Click the red triangle next to Response Defects and select **Save Columns > Predicted Values**.

14. Return to the Grill.jmp data table.

15. Change the column name for *Pred Defects* to *Poisson*.

16. Change the column name for *Predicted Defects* to *Normal*.

17. Select **Graph > Overlay Plot**.

18. Select *Defects* for **X**.

19. Select *Poisson, Normal,* and *Defects* for **Y**.

20. Click **OK**.

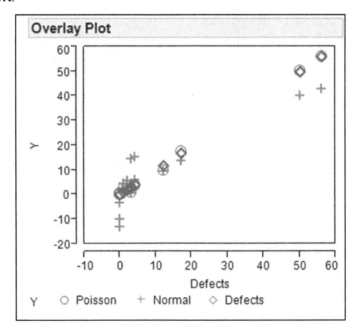

We can see that the Poisson generalized linear model provides a better fit to the data. It is particularly troublesome that the ordinary least squares (Normal) estimates are poor at the lower values for defects and are even negative in some cases.

21. Select **Window > Close All**.

Example 15.4 The Worsted Yarn Experiment

1. Open Worsted-Yarn.jmp.

2. Select **Analyze > Fit Model**.

3. Select *ln(cycles)* for **Y**.

4. Select *x1, x2,* and *x3* under **Select Columns**.

5. Click **Add**.

6. Check **Keep dialog open**.

7. Click **Run**.

8. Click the red triangle next to Response ln(cycles) and select **Estimates > Show Prediction Expression**.

Prediction Expression
6.33466432677778
+ 0.832384162111111 * x1
+ -0.6309915665 * x2
+ -0.3924939531667 * x3

This expression produces the predicted natural logarithm of the number of cycles. This data may also be fit with non-normal distributions in a generalized linear model. The textbook fits this data with a gamma distribution (a more general form of an exponential distribution) with a log link function.

9. Return to the **Fit Model** dialog.

10. Delete *ln(cycles)* from **Y** and replace it with *cycles*.

11. From the **Personality** drop-down menu, select **Generalized Linear Model**.

12. Select **Exponential** for Distribution.

13. For **Link Function**, choose **Log**.

14. Click **Run**.

Parameter Estimates

Term	Estimate	Std Error	L-R ChiSquare	Prob>ChiSq	Lower CL	Upper CL
Intercept	6.3489123	0.1924501	30484.159	<.0001*	5.9940121	6.7513875
x1	0.84251	0.2389413	11.79558	0.0006*	0.3701106	1.3148326
x2	-0.631321	0.2353273	6.9627541	0.0083*	-1.096567	-0.165803
x3	-0.385134	0.2386099	2.5764765	0.1085	-0.857011	0.0864481

These estimates result in the model found in the textbook on page 651.

15. Select **Window > Close All**.

Section 15.2 Unbalanced Data in a Factorial Design

1. Open Battery-Design.jmp.

2. Select **Analyze > Fit Model**.

3. Select *Y* for **Y**.

4. Select *Material* under **Select Columns** and click **Add.** We need to add the variables in this order because the test in this example is a sequential test.

5. Select *Temperature* under **Select Columns** and click **Add.**

6. Select *Temperature* and *Material* under **Select Columns**.

7. Click **Cross.**

8. Click **Run.**

9. Click the red triangle next to Response Y and select **Estimates > Sequential Tests**. This test (using Type I Sums of Squares) considers factors in the order they are listed. In this case, the test for the significance of *Material* does not control for *Temperature* or *Temperature*Material*. The tests produced by default with **Fit Model** are Type III Sums of Squares, which are generally more flexible and useful. Type I and Type III Sums of Squares will be the same if the design is balanced and orthogonal.

Analysis of Variance

| | | Sum of | | |
Source	DF	Squares	Mean Square	F Ratio
Model	8	30169.000	3771.13	4.6189
Error	11	8981.000	816.45	Prob > F
C. Total	19	39150.000		0.0110*

Sequential (Type 1) Tests

Source	Nparm	DF	Seq SS	F Ratio	Prob > F
Material	2	2	7811.600	4.7839	0.0320*
Temperature	2	2	16090.875	9.8541	0.0035*
Temperature*Material	4	4	6266.525	1.9188	0.1774

Both *Material* and *Temperature* are significant at the 0.05 level.

10. Select **Window > Close All**.

Example 15.5 Analysis of Covariance

1. Open Breaking-Strength.jmp.

2. Select **Analyze > Fit Model**.

3. Select *Strength* for **Y**.

4. Select *Machine* and *Diameter* under **Select Columns**. Click **Add**.

5. Click **Run**.

Analysis of Variance

| | | Sum of | | |
Source	DF	Squares	Mean Square	F Ratio
Model	3	318.41411	106.138	41.7181
Error	11	27.98589	2.544	Prob > F
C. Total	14	346.40000		<.0001*

Effect Tests

Source	Nparm	DF	Sum of Squares	F Ratio	Prob > F
Machine	2	2	13.28385	2.6106	0.1181
Diameter	1	1	178.01411	69.9694	<.0001*

Parameter Estimates

Term	Estimate	Std Error	t Ratio	Prob>|t|
Intercept	17.177096	2.783007	6.17	<.0001*
Machine[1]	0.1824131	0.594997	0.31	0.7649
Machine[2]	1.2192229	0.620117	1.97	0.0750
Diameter	0.9539877	0.114048	8.36	<.0001*

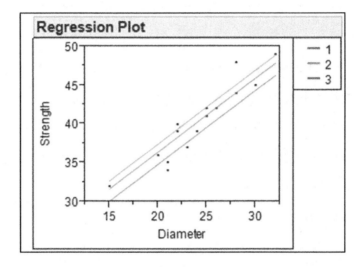

After removing the linear effect of *Diameter* on *Strength,* there is no apparent machine-to-machine variation in *Strength*. This conclusion assumes that the machines are not systematically producing different diameters of fibers. Reducing the variation in the *Diameter* of fibers produced by each machine would help reduce the overall variability in *Strength.*

6.   Select **Window > Close All**.

Section 15.3.4 Factorial Experiments with Covariates

1. Open Factorial-Covariate.jmp.

2. Select **Analyze > Fit Model**.

3. Select *y* for **Y**.

4. Select *A, B,* and *C* under **Select Columns**.

5. Select **Macros > Full Factorial**.

6. Check **Keep dialog open**.

7. Click **Run**.

Summary of Fit

RSquare	0.856776
RSquare Adj	0.731454
Root Mean Square Error	19.84247
Mean of Response	25.02875
Observations (or Sum Wgts)	16

Analysis of Variance

Source	DF	Sum of Squares	Mean Square	F Ratio
Model	7	18842.193	2691.74	6.8366
Error	8	3149.788	393.72	Prob > F
C. Total	15	21991.981		0.0073*

Parameter Estimates

Term	Estimate	Std Error	t Ratio	Prob>|t|
Intercept	25.02875	4.960616	5.05	0.0010*
A	11.19875	4.960616	2.26	0.0539
B	18.055	4.960616	3.64	0.0066*
A*B	-18.905	4.960616	-3.81	0.0052*
C	7.2425	4.960616	1.46	0.1824
A*C	14.7975	4.960616	2.98	0.0175*
B*C	-9.02625	4.960616	-1.82	0.1063
A*B*C	3.99375	4.960616	0.81	0.4440

The MSE for this model is 393.72. We will compare this to the MSE of a more complex model with an added covariate. Note that the model result shown on page 668 reduces this model by eliminating the *BC* and *ABC* higher order terms with p-values exceeding 0.10. This results in an R^2 of 0.786.

8. Return to the **Fit Model** dialog.

9. Select *x* under **Select Columns** and click **Add**.

10. Click **Run**.

Summary of Fit

RSquare	0.971437
RSquare Adj	0.938795
Root Mean Square Error	9.47287
Mean of Response	25.02875
Observations (or Sum Wgts)	16

Analysis of Variance

Source	DF	Sum of Squares	Mean Square	F Ratio
Model	8	21363.834	2670.48	29.7595
Error	7	628.147	89.74	Prob > F
C. Total	15	21991.981		<.0001*

Parameter Estimates

Term	Estimate	Std Error	t Ratio	Prob>|t|
Intercept	-1.015872	5.454106	-0.19	0.8575
A	9.4566966	2.39091	3.96	0.0055*
B	16.128277	2.395946	6.73	0.0003*
A*B	-15.59941	2.448939	-6.37	0.0004*
C	2.4287693	2.536347	0.96	0.3702
A*C	-0.419306	3.72135	-0.11	0.9135
B*C	-0.863837	2.824779	-0.31	0.7686
A*B*C	1.469927	2.4156	0.61	0.5621
x	4.9245327	0.928977	5.30	0.0011*